Einstein metrics and Yang-Mills connections

PURE AND APPLIED MATHEMATICS

A Program of Monographs, Textbooks, and Lecture Notes

LECTURE NOTES IN PURE AND APPLIED MATHEMATICS

1. *N. Jacobson,* Exceptional Lie Algebras
2. *L.-Å. Lindahl and F. Poulsen,* Thin Sets in Harmonic Analysis
3. *I. Satake,* Classification Theory of Semi-Simple Algebraic Groups
4. *F. Hirzebruch, W. D. Newmann, and S. S. Koh, Differentiable Manifolds and Quadratic Forms*
5. *I. Chavel,* Riemannian Symmetric Spaces of Rank One
6. *R. B. Burckel,* Characterization of C(X) Among Its Subalgebras
7. *B. R. McDonald, A. R. Magid, and K. C. Smith,* Ring Theory: Proceedings of the Oklahoma Conference
8. *Y.-T. Siu,* Techniques of Extension on Analytic Objects
9. *S. R. Caradus, W. E. Pfaffenberger, and B. Yood,* Calkin Algebras and Algebras of Operators on Banach Spaces
10. *E. O. Roxin, P.-T. Liu, and R. L. Sternberg,* Differential Games and Control Theory
11. *M. Orzech and C. Small,* The Brauer Group of Commutative Rings
12. *S. Thomier,* Topology and Its Applications
13. *J. M. Lopez and K. A. Ross,* Sidon Sets
14. *W. W. Comfort and S. Negrepontis,* Continuous Pseudometrics
15. *K. McKennon and J. M. Robertson,* Locally Convex Spaces
16. *M. Carmeli and S. Malin,* Representations of the Rotation and Lorentz Groups: An Introduction
17. *G. B. Seligman,* Rational Methods in Lie Algebras
18. *D. G. de Figueiredo,* Functional Analysis: Proceedings of the Brazilian Mathematical Society Symposium
19. *L. Cesari, R. Kannan, and J. D. Schuur,* Nonlinear Functional Analysis and Differential Equations: Proceedings of the Michigan State University Conference
20. *J. J. Schäffer,* Geometry of Spheres in Normed Spaces
21. *K. Yano and M. Kon,* Anti-Invariant Submanifolds
22. *W. V. Vasconcelos,* The Rings of Dimension Two
23. *R. E. Chandler,* Hausdorff Compactifications
24. *S. P. Franklin and B. V. S. Thomas,* Topology: Proceedings of the Memphis State University Conference
25. *S. K. Jain,* Ring Theory: Proceedings of the Ohio University Conference
26. *B. R. McDonald and R. A. Morris,* Ring Theory II: Proceedings of the Second Oklahoma Conference
27. *R. B. Mura and A. Rhemtulla,* Orderable Groups
28. *J. R. Graef,* Stability of Dynamical Systems: Theory and Applications
29. *H.-C. Wang,* Homogeneous Branch Algebras
30. *E. O. Roxin, P.-T. Liu, and R. L. Sternberg,* Differential Games and Control Theory II
31. *R. D. Porter,* Introduction to Fibre Bundles
32. *M. Altman,* Contractors and Contractor Directions Theory and Applications
33. *J. S. Golan,* Decomposition and Dimension in Module Categories
34. *G. Fairweather,* Finite Element Galerkin Methods for Differential Equations
35. *J. D. Sally,* Numbers of Generators of Ideals in Local Rings
36. *S. S. Miller,* Complex Analysis: Proceedings of the S.U.N.Y. Brockport Conference
37. *R. Gordon,* Representation Theory of Algebras: Proceedings of the Philadelphia Conference
38. *M. Goto and F. D. Grosshans,* Semisimple Lie Algebras
39. *A. I. Arruda, N. C. A. da Costa, and R. Chuaqui,* Mathematical Logic: Proceedings of the First Brazilian Conference
40. *F. Van Oystaeyen,* Ring Theory: Proceedings of the 1977 Antwerp Conference
41. *F. Van Oystaeyen and A. Verschoren,* Reflectors and Localization: Application to Sheaf Theory
42. *M. Satyanarayana,* Positively Ordered Semigroups
43. *D. L Russell,* Mathematics of Finite-Dimensional Control Systems
44. *P.-T. Liu and E. Roxin,* Differential Games and Control Theory III: Proceedings of the Third Kingston Conference, Part A
45. *A. Geramita and J. Seberry,* Orthogonal Designs: Quadratic Forms and Hadamard Matrices
46. *J. Cigler, V. Losert, and P. Michor,* Banach Modules and Functors on Categories of Banach Spaces

47. *P.-T. Liu and J. G. Sutinen,* Control Theory in Mathematical Economics: Proceedings of the Third Kingston Conference, Part B
48. *C. Byrnes,* Partial Differential Equations and Geometry
49. *G. Klambauer,* Problems and Propositions in Analysis
50. *J. Knopfmacher,* Analytic Arithmetic of Algebraic Function Fields
51. *F. Van Oystaeyen,* Ring Theory: Proceedings of the 1978 Antwerp Conference
52. *B. Kadem,* Binary Time Series
53. *J. Barros-Neto and R. A. Artino,* Hypoelliptic Boundary-Value Problems
54. *R. L. Sternberg, A. J. Kalinowski, and J. S. Papadakis,* Nonlinear Partial Differential Equations in Engineering and Applied Science
55. *B. R. McDonald,* Ring Theory and Algebra III: Proceedngs of the Third Oklahoma Conference
56. *J. S. Golan,* Structure Sheaves Over a Noncommutative Ring
57. *T. V. Narayana, J. G. Williams, and R. M. Mathsen,* Combinatorics, Representation Theory and Statistical Methods in Groups: YOUNG [illegible] Proceedings
58. *T. A. Burton,* Modeling and Differential Equations in Biology
59. *K. H. Kim and F. W. Roush,* Introduction to Mathematical Consensus Theory
60. *J. Banas and K. Goebel,* Measures of Noncompactness in Banach Spaces
61. *O. A. Nielson,* Direct Integral Theory
62. *J. E. Smith, G. O. Kenny, and R. N. Ball,* Ordered Groups: Proceedings of the Boise State Conference
63. *J. Cronin,* Mathematics of Cell Electrophysiology
64. *J. W. Brewer,* Power Series Over Commutative Rings
65. *P. K. Kamthan and M. Gupta,* Sequence Spaces and Series
66. *T. G. McLaughlin,* Regressive Sets and the Theory of Isols
67. *T. L. Herdman, S. M. Rankin III, and H. W. Stech,* Integral and Functional Differential Equations
68. *R. Draper,* Commutative Algebra: Analytic Methods
69. *W. G. McKay and J. Patera,* Tables of Dimensions, Indices, and Branching Rules for Representations of Simple Lie Algebras
70. *R. L. Devaney and Z. H. Nitecki,* Classical Mechanics and Dynamical Systems
71. *J. Van Geel,* Places and Valuations in Noncommutative Ring Theory
72. *C. Faith,* Injective Modules and Injective Quotient Rings
73. *A. Fiacco,* Mathematical Programming with Data Perturbations I
74. *P. Schultz, C. Praeger, and R. Sullivan,* Algebraic Structures and Applications: Proceedings of the First Western Australian Conference on Algebra
75. *L Bican, T. Kepka, and P. Nemec,* Rings, Modules, and Preradicals
76. *D. C. Kay and M. Breen,* Convexity and Related Combinatorial Geometry: Proceedings of the Second University of Oklahoma Conference
77. *P. Fletcher and W. F. Lindgren,* Quasi-Uniform Spaces
78. *C.-C. Yang,* Factorization Theory of Meromorphic Functions
79. *O. Taussky,* Ternary Quadratic Forms and Norms
80. *S. P. Singh and J. H. Burry,* Nonlinear Analysis and Applications
81. *K. B. Hannsgen, T. L. Herdman, H. W. Stech, and R. L. Wheeler,* Volterra and Functional Differential Equations
82. *N. L. Johnson, M. J. Kallaher, and C. T. Long,* Finite Geometries: Proceedings of a Conference in Honor of T. G. Ostrom
83. *G. I. Zapata,* Functional Analysis, Holomorphy, and Approximation Theory
84. *S. Greco and G. Valla,* Commutative Algebra: Proceedings of the Trento Conference
85. *A. V. Fiacco,* Mathematical Programming with Data Perturbations II
86. *J.-B. Hiriart-Urruty, W. Oettli, and J. Stoer,* Optimization: Theory and Algorithms
87. *A. Figa Talamanca and M. A. Picardello,* Harmonic Analysis on Free Groups
88. *M. Harada,* Factor Categories with Applications to Direct Decomposition of Modules
89. *V. I. Istrăţescu,* Strict Convexity and Complex Strict Convexity
90. *V. Lakshmikantham,* Trends in Theory and Practice of Nonlinear Differential Equations
91. *H. L. Manocha and J. B. Srivastava,* Algebra and Its Applications
92. *D. V. Chudnovsky and G. V. Chudnovsky,* Classical and Quantum Models and Arithmetic Problems
93. *J. W. Longley,* Least Squares Computations Using Orthogonalization Methods
94. *L. P. de Alcantara,* Mathematical Logic and Formal Systems
95. *C. E. Aull,* Rings of Continuous Functions
96. *R. Chuaqui,* Analysis, Geometry, and Probability
97. *L. Fuchs and L. Salce,* Modules Over Valuation Domains

98. *P. Fischer and W. R. Smith,* Chaos, Fractals, and Dynamics
99. *W. B. Powell and C. Tsinakis,* Ordered Algebraic Structures
100. *G. M. Rassias and T. M. Rassias,* Differential Geometry, Calculus of Variations, and Their Applications
101. *R.-E. Hoffmann and K. H. Hofmann,* Continuous Lattices and Their Applications
102. *J. H. Lightbourne III and S. M. Rankin III,* Physical Mathematics and Nonlinear Partial Differential Equations
103. *C. A. Baker and L, M. Batten,* Finite Geometrics
104. *J. W. Brewer, J. W. Bunce, and F. S. Van Vleck,* Linear Systems Over Commutative Rings
105. *C. McCrory and T. Shifrin,* Geometry and Topology: Manifolds, Varieties, and Knots
106. *D. W. Kueker, E. G. K. Lopez-Escobar, and C. H. Smith,* Mathematical Logic and Theoretical Computer Science
107. *B.-L. Lin and S. Simons,* Nonlinear and Convex Analysis: Proceedings in Honor of Ky Fan
108. *S. J. Lee,* Operator Methods for Optimal Control Problems
109. *V. Lakshmikantham,* Nonlinear Analysis and Applications
110. *S. F. McCormick,* Multigrid Methods: Theory, Applications, and Supercomputing
111. *M. C. Tangora,* Computers in Algebra
112. *D. V. Chudnovsky and G. V. Chudnovsky,* Search Theory: Some Recent Developments
113. *D. V. Chudnovsky and R. D. Jenks,* Computer Algebra
114. *M. C. Tangora,* Computers in Geometry and Topology
115. *P. Nelson, V. Faber, T. A. Manteuffel, D. L. Seth, and A. B. White, Jr.,* Transport Theory, Invariant Imbedding, and Integral Equations: Proceedings in Honor of G. M. Wing's 65th Birthday
116. *P. Clément, S. Invernizzi, E. Mitidieri, and I. I. Vrabie,* Semigroup Theory and Applications
117. *J. Vinuesa,* Orthogonal Polynomials and Their Applications: Proceedings of the International Congress
118. *C. M. Dafermos, G. Ladas, and G. Papanicolaou,* Differential Equations: Proceedings of the EQUADIFF Conference
119. *E. O. Roxin,* Modern Optimal Control: A Conference in Honor of Solomon Lefschetz and Joseph P. Lasalle
120. *J. C. Díaz,* Mathematics for Large Scale Computing
121. *P. S. Milojević,* Nonlinear Functional Analysis
122. *C. Sadosky,* Analysis and Partial Differential Equations: A Collection of Papers Dedicated to Mischa Cotlar
123. *R. M. Shortt,* General Topology and Applications: Proceedings of the 1988 Northeast Conference
124. *R. Wong,* Asymptotic and Computational Analysis: Conference in Honor of Frank W. J. Olver's 65th Birthday
125. *D. V. Chudnovsky and R. D. Jenks,* Computers in Mathematics
126. *W. D. Wallis, H. Shen, W. Wei, and L. Zhu,* Combinatorial Designs and Applications
127. *S. Elaydi,* Differential Equations: Stability and Control
128. *G. Chen, E. B. Lee, W. Littman, and L. Markus,* Distributed Parameter Control Systems: New Trends and Applications
129. *W. N. Everitt,* Inequalities: Fifty Years On from Hardy, Littlewood and Pólya
130. *H. G. Kaper and M. Garbey,* Asymptotic Analysis and the Numerical Solution of Partial Differential Equations
131. *O. Arino, D. E. Axelrod, and M. Kimmel,* Mathematical Population Dynamics: Proceedings of the Second International Conference
132. *S. Coen,* Geometry and Complex Variables
133. *J. A. Goldstein, F. Kappel, and W. Schappacher,* Differential Equations with Applications in Biology, Physics, and Engineering
134. *S. J. Andima, R. Kopperman, P. R. Misra, J. Z. Reichman, and A. R. Todd,* General Topology and Applications
135. *P Clément, E. Mitidieri, B. de Pagter,* Semigroup Theory and Evolution Equations: The Second International Conference
136. *K. Jarosz,* Function Spaces
137. *J. M. Bayod, N. De Grande-De Kimpe, and J. Martínez-Maurica,* *p*-adic Functional Analysis
138. *G. A. Anastassiou,* Approximation Theory: Proceedings of the Sixth Southeastern Approximation Theorists Annual Conference
139. *R. S. Rees,* Graphs, Matrices, and Designs: Festschrift in Honor of Norman J. Pullman
140. *G. Abrams, J. Haefner, and K. M. Rangaswamy,* Methods in Module Theory

141. *G. L. Mullen and P. J.-S. Shiue*, Finite Fields, Coding Theory, and Advances in Communications and Computing
142. *M. C. Joshi and A. V. Balakrishnan*, Mathematical Theory of Control: Proceedings of the International Conference
143. *G. Komatsu and Y. Sakane*, Complex Geometry: Proceedings of the Osaka International Conference
144. *I. J. Bakelman*, Geometric Analysis and Nonlinear Partial Differential Equations
145. *T. Mabuchi and S. Mukai*, Einstein Metrics and Yang–Mills Connections: Proceedings of the 27th Taniguchi International Symposium
146. *L. Fuchs and R. Göbel*, Abelian Groups: Proceedings of the 1991 Curaçao Conference
147. *A. D. Pollington and W. Moran*, Number Theory with an Emphasis on the Markoff Spectrum

Additional Volumes in Preparation

Einstein metrics and Yang-Mills connections

proceedings of the 27th Taniguchi international symposium

edited by
Toshiki Mabuchi
Osaka University
Osaka, Japan
Shigeru Mukai
Nagoya University
Nagoya, Japan

Marcel Dekker, Inc. **New York • Basel • Hong Kong**

Library of Congress Cataloging-in-Publication Data

Einstein metrics and Yang-Mills connections / edited by Toshiki
Mabuchi, Shigeru Mukai.
p. cm. -- (Lecture notes in pure and applied mathematics ; v.
145)
"The 27th Taniguchi International Symposium "Einstein Metrics and
Yang-Mills Connections" was held in Sanda, Japan, from December 6th
through 11th, 1990"--CIP pref.
Includes bibliographical references and index.
ISBN 0-8247-9069-3
1. Hermitian structures--Congresses. 2. Kählerian structures-
-Congresses. 3. Yang-Mills theory--Congresses. I. Mabuchi,
Toshiki. II. Mukai, Shigeru.
III. International Taniguchi Symposium (27th : 1990 : Sanda-shi,
Japan) IV. Series.
QA641.E53 1993
516.3'62--dc20 98-18074
CIP

The publisher offers discounts on this book when ordered in bulk quantities. For more information, write to Special Sales/Professional Marketing at the address below.

This book is printed on acid-free paper.

MARCEL DEKKER, INC.
270 Madison Avenue, New York, New York 10016

Current printing (last digit):
10 9 8 7 6 5 4 3 2 1

PRINTED IN THE UNITED STATES OF AMERICA

Preface

The 27th Taniguchi International Symposium "Einstein Metrics and Yang-Mills Connections" was held at Sanda, Japan, from December 6th through 11th, 1990. The symposium focused on moduli spaces of various geometric objects such as conformal structures, monopoles, Einstein metrics and Yang-Mills connections. Historically, this theme has its origin in Riemann's work and was studied extensively by algebraic geometers but in recent years, the development of nonlinear partial differential equations has allowed us to obtain new aspects of moduli spaces from topological or analytic points of view.

The present volume contains 15 chapters based on the lectures and discussions given at the symposium. The topics of the proceedings cover not only Einstein metrics and Yang-Mills connections, but also various subjects related to moduli spaces ranging from Floer cohomology and Donaldson's theorem to Yamabe-type nonlinear problems, KP-equations, and Kawamata-Viehweg's vanishing theorem. All papers are in final form and no similar versions will be published elsewhere.

Topics related to conformal structures are studied by the following authors: Bahri gives a proof of the Yamabe conjecture, without the positive mass theorem, for locally conformally flat manifolds; LeBrun obtains constructions which completely characterize the compact half-conformally-flat manifolds with nonnegative scalar curvature and semi-free isometric S^1-action; Takakuwa shows the existence of infinitely many solutions of certain conformally invariant nonlinear elliptic equations.

Several chapters explore moduli spaces from topological or gauge-theoretic points of view: Braam studies the moduli space of magnetic monopoles in relation to Donaldson's polynomial invariants; Furuta obtains a variant of the Floer cohomology with an application to a generalization of Donaldson's theorem to 4-manifolds with boundary; Kotschick shows that certain simply connected spin algebraic surfaces carry spaces of harmonic spinors whose dimension is larger than the index formula for the Dirac operator predicts; Nakajima gives a hyper-Kähler isometry between the moduli space of SU(2)-monopoles and the space of equivalence classes of solutions of Nahm's equations.

We also have chapters written from algebraic-geometric points of view: Bogomolov studies the stability of vector bundles on surfaces and curves; Enoki fully extends Kawamata-Viehweg's vanishing theorem to the case of compact Kähler manifolds; Lazarsfeld generalizes, to higher dimensional cohomology, a theorem of Kempf on deformation of symmetric products of a compact Riemann surface; Mulase completely classifies all supercommutative algebras of super differential operators in terms of graded algebraic varieties and vector bundles on them.

Differential geometric studies of Yang-Mills connections or vector bundles are carried out by the following authors: Bando gives a generalization, to a certain class of Kähler manifolds, of Donaldson's characterization of the moduli space of anti-self-dual connections on the complex 2-plane with finite action; Koiso proves the existence and the mountain pass lemma for Yang-Mills connections on homogeneous bundles over noncompact homogeneous spaces.

Finally, the moduli space of compact complex manifolds is studied in two articles: Nadel treats the finiteness of the deformation types of nonsingular Fano varieties of a given dimension; Mabuchi and Mukai determine the Gromov's compactification of Einstein-Kähler quartic del Pezzo surfaces by studying their stability.

We wish to thank Mr. Toyosaburo Taniguchi and his foundation for warm hospitality and generous financial support. Special thanks are due also to Professor Shingo Murakami to whom we owe much for the organization of the symposium.

Toshiki Mabuchi
Shigeru Mukai

Contents

Contributors

Abbas Bahri:

Department of Mathematics, Rutgers University, New Brunswick, NJ 08903, U.S.A.

Shigetoshi Bando:

Mathematical Institute, Faculty of Science, Tohoku University, Sendai, 980 Japan

Fedor Bogomolov:

Department of Algebra, Steklov Institute of Mathematics, ul. Vavilova 42, GSP-1, 117966 Moscow, Russia

Peter Braam:

Mathematical Institute, University of Oxford, 24-29 St Giles', Oxford, OX1 3LB, England

Ichiro Enoki:

Department of Mathematics, College of General Education, Osaka University, Toyonaka, Osaka, 560 Japan

Mikio Furuta:

Department of Mathematical Sciences, University of Tokyo, Komaba, Meguro-ku, Tokyo, 153 Japan

Norihito Koiso:

Department of Mathematics, College of General Education, Osaka University, Toyonaka, Osaka, 560 Japan

Dieter Kotschick:

Queens' College, Cambridge, CB3 9ET, England

Robert Lazarsfeld:

Department of Mathematics, University of California, Los Angeles, CA 90024, U.S.A.

Claude LeBrun:

Department of Mathematics, State University of New York, Stony Brook, NY 11794-3651, U.S.A.

Toshiki Mabuchi:

Department of Mathematics, College of General Education, Osaka University, Toyonaka, Osaka, 560 Japan

Shigeru Mukai:

Department of Mathematics, Faculty of Science, Nagoya University, Chikusa-ku, Nagoya, 464-01 Japan

Motohico Mulase:

Department of Mathematics, University of California, Davis, CA 95616, U.S.A.

Alan Nadel:

School of Mathematics, The Institute for Advanced Study, Princeton, NJ 08540, U.S.A.

Hiraku Nakajima:*

Department of Mathematics, University of Tokyo, Hongo 7-3-1, Tokyo, 113 Japan

Shoichiro Takakuwa:

Department of Mathematics, Faculty of Science, Tokyo Metropolitan University, Hachi-oji, Tokyo, 192-03 Japan

*Current Address: Mathematical Insititute, Tôhoku University, Aramaki, Aoba-ku, Sendai 980, Japan

Einstein metrics and Yang-Mills connections

1

Proof of the Yamabe Conjecture, Without the Positive Mass Theorem, for Locally Conformally Flat Manifolds

ABBAS BAHRI*

Rutgers University and Ecole Nationale d'Ingénieurs de Tunis

Let (M^n, g) be a compact locally conformally flat manifold of dimension $n \geq 3$. Let $L = \Delta_g - \frac{(n-2)}{4(n-1)}R(g)$ be the conformal Laplacian on M. The Yamabe conjecture for such manifolds has been established by T. Aubin [**1**] and R. Schoèn [**2**]. This conjecture is equivalent to the existence of a positive function u such that:

$$(1) \qquad \begin{cases} -Lu = u^{n+2/n-2} \\ u > 0 \ \text{ on } M \end{cases}$$

(1) falls in the framework of a more general class of equations of the type:

$$(2) \qquad \begin{cases} -Lu + qu = u^{n+2/n-2} \\ q \in L^\infty(M) \\ u > 0 \ \text{ on } M \end{cases}$$

under the assumption that the first eigenvalue $\lambda_1(-L+q)$ of $-L+q$ is positive. Another equation close to (1) is the following equation:

$$(3) \qquad \begin{cases} -\Delta u = u^{n+2/n-2} \\ u > 0, \ \ \Omega : \text{bounded regular} \subset \mathbb{R}^n \\ u_{|\partial\Omega} = 0. \end{cases}$$

*The author wishes to thank Mrs. Henda El Fekih, from the Ecole Nationale d'Ingénieurs de Tunis for her kind assistance during the preparation of this manuscript.

While the proof of the Yamabe conjecture by R. Schoèn [**2**] involves the positive mass theorem [**3**], equations of the type (2) and (3) have also been studied through another method, of topological type, involving the study of the "critical points at infinity" ([**4**], [**5**], [**6**], [**7**]). In [**7**], we established that (2) had a solution under various assumptions on M. There were no assumptions if $\dim M = 3, 4, 5$. In [**6**], we established that (3) had a solution under topological assumptions on Ω.

We prove here the following theorem, using the same technics. This theorem together with the results of [**6**] and [**7**], provides a proof of the Yamabe conjecture based on the method of "critical points at infinity," which does not use the positive mass theorem. We thus have

Theorem 1. *Assume $n \geq 3$. Then the equation* (1) *has a solution.*

The proof of Theorem 1 involves several steps which are detailed below. We first need the following construction: since M is compact and locally conformally flat, any point x_0 of M has a neighborhood $U(x_0) \supset B(x_0, \rho)$, $\rho > 0$ uniform, such that the metric g on $U(x_0)$ is conformal to an euclidian metric $\bar{g}_0$. We may assume that the euclidian metric $\bar{g}_0$ on $U(x_0)$ depends smoothly on x_0 (see Appendix C). We may therefore find a positive function $\tilde{u}_{x_0}$ on $B(x_0, \rho)$ such that:

$$\tilde{u}_{x_0}^{4/n-2} g = \bar{g}_0. \tag{4}$$

Let u_{x_0} be the following function:

$$u_{x_0}(x) = \tilde{u}_{x_0}(x) \qquad \text{on } B(x_0, \rho/2) \tag{5}$$

$$u_{x_0}(x) = \omega_{x_0}\tilde{u}_{x_0} + (1 - \omega_{x_0}) \qquad \text{on } B(x_0, \rho) - B(x_0, \rho/2) \tag{6}$$

where ω_{x_0} is a C^∞ function, valued in $[0, 1]$, equal to 1 on $B(x_0, \rho/2)$ and equal to zero on $M - B(x_0, \rho)$. $u_{x_0}(x)$ is then a C^∞ function, which depends smoothly on x_0.

Let λ be a positive, large parameter. We define:

$$(7) \qquad \delta(x_0,\lambda)(x) = \bar{c}\frac{\lambda^{n-2/2}}{(1+\lambda^2\bar{g}_0(x-x_0,x-x_0))^{n-2/2}} \qquad \text{if } x \in B(x_0,\rho).$$

$\bar{c}$ is normalized so that the following equation is satisfied:

$$(8) \qquad -\Delta_{\bar{g}_0}\delta = \delta^{n+2/n-2} \qquad \text{on } B(x_0,\rho).$$

Observe that, since $\bar{g}_0$ is euclidian, the scalar curvature $R(\bar{g}_0)$ is zero and $-\Delta_{\bar{g}_0}$ is the conformal Laplacian $-L_0$ of the metric $\bar{g}_0$. We may think of $B(x_0,\rho)$ as a ball in $\mathbb{R}^n$ and of $\bar{g}_0$ as a scalar product on $\mathbb{R}^n$. Then, formula (7) can be extended to all of $\mathbb{R}^n$. In the sequel, we denote also $\delta(x_0,\lambda)$ this extension.

We define now: a family of functions $\hat{\delta}(x_0,\lambda)$, depending on $x_0 \in M$ and $\lambda > 0$, λ large. These functions are being used here, in the existence proof, in the same way the functions $\hat{\delta}$ are used in [**7**], or the functions $P\delta$ in [**6**]. The formula for $\hat{\delta}$ is:

$$(9) \quad -L\hat{\delta}(x_0,\lambda)(x) = \begin{cases} (\,\omega_{x_0}(x)u_{x_0}(x)\delta(x_0,\lambda)(x)\,)^{n+2/n-2} & \text{in } B(x_0,\rho) \\ 0 & \text{in } M - B(x_0,\rho). \end{cases}$$

We introduce also the functions $\delta'(x_0,\lambda)$, defined by the formula:

$$(10) \qquad \delta'(x_0,\lambda)(x) = \begin{cases} \omega_{x_0}(x)u_{x_0}(x)\delta(x_0,\lambda)(x) & \text{in } B(x_0,\rho) \\ 0 & \text{in } M - B(x_0,\rho). \end{cases}$$

We need now the following lemmas:

Lemma 2. *Let $\Psi \in C^2(B(x_0, \rho/2), \mathbb{R})$. The following formula holds:*

$$-L(u_{x_0}(x)\Psi) = u_{x_0}^{n+2/n-2}(-L_0\Psi).$$

Lemma 3. *There exist a constant $C > 0$ and a constant $B > 0$ such that, for any (x_0, λ), $x_0 \in M$ and $\lambda \geq B$, the following estimate holds:*

$$|\hat{\delta}(x_0, \lambda) - \delta'(x_0, \lambda)|_\infty \leq \frac{C}{\lambda^{n-2/2}}.$$

Proof of Lemma 2:

Since g and $\bar{g}_0$ are conformal, with a factor equal to $u_{x_0}^{4/n-2}$, we have, for any $\Psi_1 \in C_0^2(B(x_0, \rho/2), \mathbb{R})$:

$$\nabla_g \Psi \nabla_g \Psi_1 = u_{x_0}^{4/n-2} \nabla_{\bar{g}_0} \Psi \nabla_{\bar{g}_0} \Psi_1 \tag{11}$$

hence, since $dv_{\bar{g}_0} = u_{x_0}^{2n/n-2} dv_g$:

$$u_{x_0}^2 \nabla_g \Psi \nabla_g \Psi_1 dv_g = \nabla_{\bar{g}_0} \Psi \nabla_{\bar{g}_0} \Psi_1 dv_{\bar{g}_0}. \tag{12}$$

Integrating, we derive:

$$-\int_{B(x_0,\rho/2)} \nabla^g (u_{x_0}^2 \nabla_g \Psi) \Psi_1 dv_g = \int_{B(x_0,\rho/2)} (-L_0 \Psi) \Psi_1 dv_{\bar{g}_0}. \tag{13}$$

Thus:

$$\begin{aligned}
-L_0\Psi &= -u_{x_0}^{-2n/n-2} \nabla^g (u_{x_0}^2 \nabla_g \Psi) \\
&= -u_{x_0}^{n+2/2-n} \nabla^g (u_{x_0} \nabla_g \Psi) - u_{x_0}^{n+2/2-n} \nabla^g \Psi \cdot \nabla_g u_{x_0} \\
&= -u_{x_0}^{n+2/2-n} \nabla^g (\nabla_g (u_{x_0} \Psi)) + u_{x_0}^{n+2/2-n} \nabla^g (\Psi \nabla_g u_{x_0}) \\
&\quad - u_{x_0}^{n+2/2-n} \nabla^g \Psi \cdot \nabla_g u_{x_0} \\
&= -u_{x_0}^{n+2/2-n} \left\{ \nabla^g (\nabla_g (u_{x_0} \Psi)) - \frac{(n-2)}{4(n-1)} R(g) u_{x_0} \Psi \right\} \\
&\quad + u_{x_0}^{n+2/2-n} \left\{ (\Delta_g u_{x_0}) \Psi - \frac{(n-2)}{4(n-1)} R(g) u_{x_0} \Psi \right\}.
\end{aligned} \tag{14}$$

Since $\bar{g}_0$ is flat, we have:

$$\Delta_g u_{x_0} - \frac{(n-2)}{4(n-1)} R(g) u_{x_0} = 0 \tag{15}$$

hence

$$-L_0 \Psi = -u_{x_0}^{n+2/2-n} L(u_{x_0} \Psi).$$

Lemma 2 is thereby proven.

Proof of Lemma 3:

We have on $B(x_0, \rho/2)$:

$$\begin{aligned} -L\delta'(x_0,\lambda)(x) &= -L(u_{x_0}\delta(x_0,\lambda)) = -u_{x_0}^{n+2/n-2} L_0 \delta(x_0,\lambda) \\ &= u_{x_0}^{n+2/n-2} \delta(x_0,\lambda)^{n+2/n-2} = -L\hat{\delta}(x_0,\lambda)(x). \end{aligned} \tag{16}$$

On the other hand, we may find a constant C, depending on ρ and not on x_0 such that:

$$|-L\hat{\delta}(x_0,\lambda)| + |-L\delta'(x_0,\lambda)| \le \frac{C}{\lambda^{n-2/2}} \quad \text{if } x \in M - B(x_0, \rho/2) \tag{17}$$

provided λ is large enough ($\lambda \ge 1$ will suffice). Thus

$$|-L((\hat{\delta} - \delta')(x_0,\lambda)(x))| \le \frac{C}{\lambda^{n-2/2}}. \tag{18}$$

Using the maximum principle, Lemma 3 follows.

The next step in the proof of Theorem 1 is provided by the following lemma:

Lemma 4. *Let $\theta > 0$ be given. There are positive constants C and B such that the following estimates hold, provided $\lambda \geq B$:*

$$\text{(i)} \qquad \hat{\delta}(x_0,\lambda)(x) \geq \frac{1}{C\lambda^{n-2/2}} \qquad \forall x_0 \in M;$$

$$\text{(ii)} \qquad \left| -\int_M L\hat{\delta}(x_0,\lambda)\hat{\delta}(x_0,\lambda)dv_g - \int_{\mathbf{R}^n} \bar{c}^2 \left|\nabla(\frac{1}{(1+|x|^2)^{n-2/2}})\right|^2 dx \right| \leq \frac{C}{\lambda^{n-2}} \quad \forall x_0 \in M;$$

$$\text{(iii)} \qquad \left| \int_M \hat{\delta}(x_0,\lambda)^{2n/n-2} dv_g - \int_{\mathbf{R}^n} \bar{c}^{2n/n-2} \frac{1}{(1+|x|^2)^n} dx \right| \leq \frac{C}{\lambda^{n-2}} \quad \forall x_0 \in M;$$

$$\text{(iv)} \qquad -\int_M L\hat{\delta}(x_0,\lambda)\hat{\delta}(x_1,\lambda)dv_g \leq (1+\theta)\int_M \hat{\delta}(x_0,\lambda)^{n+2/n-2}\hat{\delta}(x_1,\lambda)dv_g \quad \forall x_0, x_1 \in M;$$

$$\text{(v)} \qquad \int_M \hat{\delta}(x_0,\lambda)^{n+2/n-2}\hat{\delta}(x_1,\lambda)dv_g \geq \frac{1}{C\lambda^{n-2}}$$

Proof of Lemma 4:

Proof of (i):

Using Lemma 3, we know that if $\rho_1 < \rho$ is chosen small enough, independent of λ, the following inequality holds on $B(x_0, \rho_1)$:

$$\hat{\delta}(x_0,\lambda)(x) \geq \delta'(x_0,\lambda)(x) - \frac{C}{\lambda^{n-2/2}} \geq \frac{C'}{\lambda^{n-2/2}} \quad \forall x \in B(x_0,\rho_1) \tag{19}$$

where $C' > 0$ is independent of λ. Thus, we have:

$$\begin{cases} -L\hat{\delta}(x_0,\lambda) \geq 0 & \text{in } M - B(x_0,\rho_1), \\ \hat{\delta}(x_0,\lambda)_{|\partial B(x_0,\rho_1)} \geq \frac{C'}{\lambda^{n-2/2}}. \end{cases} \tag{20}$$

(i) follows using the maximum principle.

Proof of (ii):

We have:

$$(21)\quad -\int_M L\hat{\delta}(x_0,\lambda)\hat{\delta}(x_0,\lambda)dv_g = \int_{B(x_0,\rho)} \{\omega_{x_0}(x)\delta(x_0,\lambda)u_{x_0}\}^{n+2/n-2}\hat{\delta}(x_0,\lambda)dv_g$$

hence, using Lemma 3:

$$(22)\quad \begin{aligned} & -\int_M L\hat{\delta}(x_0,\lambda)\hat{\delta}(x_0,\lambda)dv_g \\ &= \int_{B(x_0,\rho)} \omega_{x_0}(x)^{2n/n-2}\delta(x_0,\lambda)^{2n/n-2}u_{x_0}(x)^{2n/n-2}dv_g \\ &\quad + O(\frac{1}{\lambda^{n-2/2}})\int_{B(x_0,\rho)} \delta(x_0,\lambda)^{n+2/n-2}dv_g \\ &= \int_{B(x_0,\rho)} \omega_{x_0}(x)^{2n/n-2}\delta(x_0,\lambda)^{2n/n-2}dv_{\bar{g}_0} \\ &\quad + O(\frac{1}{\lambda^{n-2/2}})\int_{B(x_0,\rho)} \delta(x_0,\lambda)^{n+2/n-2}dv_{\bar{g}_0} \\ &= \int_{B(x_0,\rho)} \delta(x_0,\lambda)^{2n/n-2}dv_{\bar{g}_0} + O(\frac{1}{\lambda^{n-2}}) \\ &= \int_{\mathbf{R}^n} \delta(x_0,\lambda)^{2n/n-2}dv_{\bar{g}_0} + O(\frac{1}{\lambda^{n-2}}) \\ &= \int_{\mathbf{R}^n} \bar{c}^2\left|\nabla(\frac{1}{(1+|x|^2)^{n-2/2}})\right|^2 dx + O(\frac{1}{\lambda^{n-2}}). \end{aligned}$$

In (22), we used various easy estimates established in [**5**], [**6**],[**7**], in particular, such as

$$\int_{B(x_0,\rho)-B(x_0,\rho/2)} \delta(x_0,\lambda)^{2n/n-2}\, dv_{\bar{g}_0} = O(\frac{1}{\lambda^{n-2}}),$$

$u_{x_0} \geq c > 0$ on $B(x_0,\rho)$ (obviously true), $\int_{B(x_0,\rho)} \delta(x_0,\lambda)^{n+2/n-2}dv_{\bar{g}_0} = O(\frac{1}{\lambda^{n-2/2}})$ etc. (ii) follows from (22).

Proof of (iii):

Using Lemma 3, the proof of (iii) essentially reduces to the computations in (22), up to minor differences.

Proof of (v):

Using (i), we derive:

$$\int_M \hat\delta(x_0,\lambda)^{n+2/n-2}\hat\delta(x_1,\lambda)dv_g \geq \frac{1}{C\lambda^{n-2/2}}\int_{B(x_0,\rho)} \hat\delta(x_0,\lambda)^{n+2/n-2}dv_g. \tag{23}$$

Using again (i) and Lemma 3, this implies:

$$\begin{aligned}&\int_M \hat\delta(x_0,\lambda)^{n+2/n-2}\hat\delta(x_1,\lambda)dv_g \\ &\geq \frac{1}{C'\lambda^{n-2/2}}\int_{B(x_0,\rho/2)} \delta(x_0,\lambda)^{n+2/n-2}u_{x_0}^{2n/n-2}dv_g \geq \frac{1}{C\lambda^{n-2}}.\end{aligned} \tag{24}$$

Proof of (iv):

We have:

$$\begin{aligned}&-\int_M L\hat\delta(x_0,\lambda)\hat\delta(x_1,\lambda)dv_g \\ &= \int_{B(x_0,\rho)} \{\,\omega_{x_0}(x)u_{x_0}\delta(x_0,\lambda)\,\}^{n+2/n-2}\hat\delta(x_1,\lambda)dv_g \\ &= \int_{B(x_0,\rho/2)} \delta'(x_0,\lambda)^{n+2/n-2}\hat\delta(x_1,\lambda)dv_g + O(\frac{1}{\lambda^{n+2/2}})\int_{B(x_0,\rho)}\hat\delta(x_1,\lambda)dv_g \\ &= \int_{B(x_0,\rho)} \hat\delta(x_0,\lambda)^{n+2/n-2}\hat\delta(x_1,\lambda)dv_g \\ &\quad + O(\frac{1}{\lambda^{n-2/2}})\int_{B(x_0,\rho)}\delta(x_0,\lambda)^{4/n-2}\hat\delta(x_1,\lambda)dv_g \\ &\quad + O(\frac{1}{\lambda^{n+2/2}})\left(\int_{B(x_0,\rho)}\hat\delta(x_1,\lambda)^{n+2/n-2}dv_g\right)^{n-2/n+2} \\ &= \int_M \hat\delta(x_0,\lambda)^{n+2/n-2}\hat\delta(x_1,\lambda)dv_g + O(\frac{1}{\lambda^{n+2/2}}\cdot\frac{1}{\lambda^{(n-2)^2/2(n+2)}}) \\ &\quad + O(\frac{1}{\lambda^{n-2/2}})\int_{B(x_0,\rho)}\delta(x_0,\lambda)^{4/n-2}\delta(x_1,\lambda)dv_g.\end{aligned} \tag{25}$$

Observe that $\frac{n+2}{2}+\frac{(n-2)^2}{2(n+2)} > n-2$. In Appendix A, we prove that:

$$\frac{1}{\lambda^{n-2/2}}\int_{B(x_0,\rho)}\delta(x_0,\lambda)^{4/n-2}\delta(x_1,\lambda)dv_{\bar g_0} = o(\int_{\mathbf{R}^n}\delta(x_0,\lambda)^{n+2/n-2}\delta(x_1,\lambda)dv_{\bar g_0}). \tag{26}$$

Thus, using Lemma 3:

$$\frac{1}{\lambda^{n-2/2}}\int_{B(x_0,\rho)}\delta(x_0,\lambda)^{4/n-2}\delta(x_1,\lambda)dv_{\bar{g}_0} \tag{27}$$
$$= o(\int_M \hat{\delta}(x_0,\lambda)^{n+2/n-2}\hat{\delta}(x_1,\lambda)dv_g) + o(\frac{1}{\lambda^{n-2}}).$$

(25) and (27) imply:

$$-\int_M L\hat{\delta}(x_0,\lambda)\hat{\delta}(x_1,\lambda)dv_g \tag{28}$$
$$= \int_M \hat{\delta}(x_0,\lambda)^{n+2/n-2}\hat{\delta}(x_1,\lambda)dv_g + o(\int_M \hat{\delta}(x_0,\lambda)^{n+2/n-2}\hat{\delta}(x_1,\lambda)dv_g)$$
$$+ o(\frac{1}{\lambda^{n-2}}).$$

Hence, using (v):

$$-\int_M L\hat{\delta}(x_0,\lambda)\hat{\delta}(x_1,\lambda)dv_g = \int_M \hat{\delta}(x_0,\lambda)^{n+2/n-2}\hat{\delta}(x_1,\lambda)dv_g(1+o(1)). \tag{29}$$

The proof of (iv) and the proof of Lemma 4 are thereby complete.

We now introduce the functional:

$$J(u) = \frac{\left(\int_M -Lu\cdot u dv_g\right)^{n/n-2}}{\int_M |u|^{2n/n-2}dv_g} \tag{30}$$

where u belongs to the space Σ^+ defined as follows:

$$\Sigma^+ = \{\, u \in H^1(M)\,;\, u \geq 0,\, u \not\equiv 0\,\}. \tag{31}$$

Let S be defined as follows:

$$S = \left(\int_{\mathbf{R}^n}\left|\nabla\frac{1}{(1+|x|^2)^{n-2/2}}\right|^2 dx\right)^{n/n-2} \Big/ \int_{\mathbf{R}^n}\frac{dx}{(1+|x|^2)^n}\,. \tag{32}$$

We then have:

Lemma 5. (i) $\forall p \in \mathbb{N}^*$, $\forall \varepsilon_1 > 0$, $\exists \lambda_p = \lambda(p, \varepsilon_1)$ *such that for any* $(\alpha_1, \dots, \alpha_p)$ *satisfying* $\alpha_i \geq 0$, $\Sigma_{i=1}^p \alpha_i = 1$, *for any* $(x_1, \dots, x_p) \in M^p$, *for any* $\lambda \geq \lambda_p$, *the following inequality holds:*

$$J\left(\sum_{i=1}^{p} \alpha_i \hat{\delta}(x_i, \lambda)\right) \leq p^{2/n-2} S \quad \text{provided} \sum_{i \neq j} \int_M \hat{\delta}(x_i, \lambda)^{n+2/n-2} \hat{\delta}(x_j, \lambda) dv_g \geq \varepsilon_1.$$

(ii) $\exists 0 < \theta_0 < 1$, $\exists c > 0$, $\exists \bar{\varepsilon}_1 > 0$, *such that for any* $p \in \mathbb{N}^*$, *for any* $(\alpha_1, \dots, \alpha_p)$ *satisfying* $\alpha_i \geq 0$, $\Sigma_{i=1}^p \alpha_i = 1$, $\alpha_i / \alpha_j \geq \theta_0 \; \forall i, \forall j$, *for any* $(x_1, \dots, x_p) \in M^p$, *for any* $\lambda \geq 1$, *the following inequality holds:*

$$J\left(\sum_{i=1}^{p} \alpha_i \hat{\delta}(x_i, \lambda)\right) \leq \frac{(\Sigma_{i=1}^p \alpha_i^2)^{n/n-2}}{\Sigma_{i=1}^p \alpha_i^{2n/n-2}} S\left(1 + \frac{1}{c\lambda^{n-2}} - \frac{(p+1)c}{\lambda^{n-2}}\right)$$
$$\text{provided} \sum_{i \neq j} \int_M \hat{\delta}(x_i, \lambda)^{n+2/n-2} \hat{\delta}(x_j, \lambda) dv_g \leq \bar{\varepsilon}_1.$$

(iii) *If in* (ii) *we drop the condition* $\alpha_i / \alpha_j \geq \theta_0$, *then the following weaker inequality still holds:*

$$J\left(\sum_{i=1}^{p} \alpha_i \hat{\delta}(x_i, \lambda)\right)$$
$$\leq \frac{(\Sigma_{i=1}^p \alpha_i^2)^{n/n-2}}{\Sigma_{i=1}^p \alpha_i^{2n/n-2}} S \left(1 + O\left(\frac{1}{\lambda^{n-2}}\right) + \frac{1}{c} \sum_{i \neq j} \int_M \hat{\delta}(x_i, \lambda)^{n+2/n-2} \hat{\delta}(x_j, \lambda) dv_g\right).$$

Lemma 5 implies the following proposition:

Proposition 6. *There exists an integer* $p_0 > 0$ *and a positive real number* $\lambda_0 > 0$ *such that for any* $(\alpha_1, \dots, \alpha_p)$ *satisfying,* $\alpha_i \geq 0$, $\Sigma_{i=1}^p \alpha_i = 1$, *for any* $(x_1, \dots, x_p) \in M^p$, *for any* $\lambda \geq \lambda_0$, *we have:*

$$J\left(\sum_{i=1}^{p_0} \alpha_i \hat{\delta}(x_i, \lambda)\right) \leq p_0^{2/n-2} S.$$

Proof of Proposition 6:

The proof of Proposition 6 follows easily from (i), (ii) and (iii) of Lemma 5: We first choose $0 < \varepsilon_1 < \bar{\varepsilon}_1$ and λ_0 so that:

$$\frac{(\Sigma_{i=1}^p \alpha_i^2)^{n/n-2}}{\Sigma_{i=1}^p \alpha_i^{2n/n-2}} \cdot \left\{1 + O(\frac{1}{\lambda^{n-2}}) + \frac{\varepsilon_1}{c}\right\} < 1$$

if $\lambda \geq \lambda_0$ and if $\alpha_{i_0}/\alpha_{j_0} \leq \theta_0 < 1$ for a couple of indexes (i_0, j_0).

Considering $(\alpha_1, \ldots, \alpha_p)$, $(x_1, \ldots, x_p)$ and $\lambda \geq \lambda_0$, we study various cases: either $\alpha_{i_0}/\alpha_{j_0} \leq \theta_0$ for a couple of indexes (i_0, j_0). Then, taking $\lambda \geq \lambda_{p_0} = \sup(\lambda(p_0, \varepsilon_1), \lambda_0)$, where $\lambda(p_0, \varepsilon_1)$ is provided by (i) of Lemma 5, we derive that:

$$J(\sum_{i=1}^p \alpha_i \hat{\delta}(x_i, \lambda)) \leq p^{2/n-2} S$$

$$\text{if } \sum_{i \neq j} \int_M \hat{\delta}(x_i, \lambda)^{n+2/n-2} \hat{\delta}(x_j, \lambda) dv_g \geq \varepsilon_1.$$

If, on the contrary, $\Sigma_{i \neq j} \int_M \hat{\delta}(x_i, \lambda)^{n+2/n-2} \hat{\delta}(x_j, \lambda) dv_g \leq \varepsilon_1$, we apply (iii). Since we choose ε_1 so that

$$\frac{(\Sigma_{i=1}^p \alpha_i^2)^{n/n-2}}{\Sigma_{i=1}^p \alpha_i^{2n/n-2}} \cdot \left\{1 + O(\frac{1}{\lambda^{n-2}}) + \frac{\varepsilon_1}{c}\right\} < 1,$$

we derive:

$$J(\sum_{i=1}^{p_0} \alpha_i \hat{\delta}(x_i, \lambda)) \leq p_0^{2/n-2} S \tag{33}$$

and Proposition 6 therefore holds in this case.

Let us assume now that $\alpha_i/\alpha_j > \theta_0$ for any (i, j). Then either (i) or (ii) of Lemma 5 holds. If (i) holds, Proposition 6 holds. We may thus assume that (ii) holds. If p_0 is chosen so that:

$$(p_0 + 1)c^2 > 1 \tag{34}$$

then

$$J\Big(\sum_{i=1}^{p_0} \alpha_i \hat{\delta}(x_i, \lambda)\Big) \leq \frac{(\Sigma_{i=1}^{p_0} \alpha_i^2)^{n/n-2}}{\Sigma_{i=1}^{p_0} \alpha_i^{2n/n-2}} S. \tag{35}$$

(33) and (35) imply Proposition 6.

Proof of Theorem 1 completed:

Proposition 6 provides the same result as Proposition B1 of [**6**], which is a key point in the existence proof completed in [**6**]. If we exclude from [**6**] Appendix B (which is replaced here by Proposition 6 as explained), the rest of [**6**] holds in this new framework, up to minor details: Propositions 6 and 7 of [**6**] hold also here and their proof requires only minor adaptations. Theorem 1 in [**6**] is derived from these propositions and Proposition B1, via a topological argument displayed in pages 260-266 of [**6**]. The same topological argument allows us to derive Theorem 1 of this present framework from the analogues of Propositions 6 and 7 of [**6**] and from Proposition 6 of the present paper. The proof of Theorem 1 is thereby complete.

Proof of Lemma 5: (the proof is extracted from [**7**])

Proof of (i):

Observe that

$$-L\hat{\delta}(x_i, \lambda) = \delta'(x_i, \lambda)^{n+2/n-2}. \tag{36}$$

Let

$$a'_i = \frac{\alpha_i \delta'(x_i,\lambda)}{\Sigma_{j=1}^p \alpha_j \delta'(x_j,\lambda)} \qquad \text{if } \delta'(x_i,\lambda)(x) \neq 0; \tag{37}$$

$$a'_i = 1 \qquad \text{if } \delta'(x_i,\lambda)(x) = 0. \tag{38}$$

Lemma B2 of [**6**] and its proof can be extended verbatim to this new framework, with $\hat{\delta}(x_i,\lambda)$ replacing $P\delta(x_i,\lambda)$ and $\delta'(x_i,\lambda)$ replacing $\delta(x_i,\lambda)$. In the proof of this lemma in [**6**], $a_i = \frac{\alpha_i\delta(x_i,\lambda)}{\Sigma_{j=1}^p \alpha_j \delta(x_j,\lambda)}$ is the global formula for a_i, while here a'_i is defined differently if $\delta'(x_i,\lambda)(x) = 0$ or $\delta'(x_i,\lambda)(x) \neq 0$. However, this fact has no incidence on the statement of the lemma or its proof, since a_i or a'_i are always multiplied by $\delta(x_i,\lambda)$ or $\delta'(x_i,\lambda)$ respectively, hence $a_i\delta(x_i,\lambda) = 0$ and $a'_i\delta'(x_i,\lambda) = 0$ if $\delta(x_i,\lambda)$ or $\delta'(x_i,\lambda)$ are zero respectively. We thus have:

$$J(\sum_{i=1}^p \alpha_i\hat{\delta}(x_i,\lambda)) \le \left\{\frac{\int_M(\sum_{i=1}^p \alpha_i\delta'(x_i,\lambda))^{2n/n-2}dv_g}{\int_M(\sum_{i=1}^p \alpha_i\hat{\delta}(x_i,\lambda))^{2n/n-2}dv_g}\right\}^{1/2} \cdot \left(\sum_{i=1}^p \int_M a'_i\delta'(x_i,\lambda)^{2n/n-2}dv_g\right)^{2/n-2}. \tag{39}$$

Using Lemma 3, we derive easily from the fact $\int_M \delta(x_i,\lambda)^{n+2/n-2}dv_g = O(\frac{1}{\lambda^{n-2/2}})$:

$$J(\sum_{i=1}^p \alpha_i\hat{\delta}(x_i,\lambda)) \le \left(1 + O(\frac{1}{\lambda^{n-2}})\right)\left(\sum_{i=1}^p \int_M a'_i\delta'(x_i,\lambda)^{2n/n-2}dv_g\right)^{2/n-2}.$$

We also have:

$$\int_M \hat{\delta}(x_i,\lambda)^{2n/n-2} = S^{n-2/2} + O(\frac{1}{\lambda^{n-2}}).$$

We thus obtain (compare with (B.41) in [**6**]):

$$J(\sum_{i=1}^p \alpha_i\hat{\delta}(x_i,\lambda)) \le \left(1 + O(\frac{1}{\lambda^{n-2}})\right) \cdot \left\{(p-1)S^{n-2/2} + \int_M \frac{\alpha_1\delta'(x_1,\lambda)\cdot\delta'(x_1,\lambda)^{2n/n-2}dv_g}{\alpha_1\delta'(x_1,\lambda) + \alpha_2\delta'(x_2,\lambda)}\right\}^{2/n-2}. \tag{40}$$

We may assume, without loss of generality, that $\alpha_1/\alpha_2 \leq 1$ and that

$$\int_M (\hat{\delta}(x_1,\lambda)^{n+2/n-2}\hat{\delta}(x_2,\lambda) + \hat{\delta}(x_2,\lambda)^{n+2/n-2}\hat{\delta}(x_1,\lambda))dv_g$$
$$= \sup\left\{\int_M (\hat{\delta}(x_i,\lambda)^{n+2/n-2}\hat{\delta}(x_j,\lambda) + \hat{\delta}(x_j,\lambda)^{n+2/n-2}\hat{\delta}(x_i,\lambda))dv_g\right\}.$$

We then have, using the fact that $\int_{M-B(x_1,\rho/2)} \delta(x_1,\lambda)^{2n/n-2}dv_g = O(\frac{1}{\lambda^{n-2}})$:

$$(41) \quad J(\sum_{i=1}^{p} \alpha_i\hat{\delta}(x_i,\lambda)) \leq$$
$$\left(1+O(\frac{1}{\lambda^{n-2}})\right)\cdot\left\{(p-1)S^{n-2/2} + \int_{B(x_1,\rho/2)} \frac{\delta'(x_1,\lambda)\cdot\delta'(x_1,\lambda)^{2n/n-2}dv_g}{\delta'(x_1,\lambda)+\delta'(x_2,\lambda)}\right\}^{2/n-2}$$

hence

$$(42) \quad J(\sum_{i=1}^{p} \alpha_i\hat{\delta}(x_i,\lambda)) \leq$$
$$\left(1+O(\frac{1}{\lambda^{n-2}})\right)\cdot\left\{(p-1)S^{n-2/2} + \int_{B(x_1,\rho/4)} \frac{\delta'(x_1,\lambda)\cdot\delta(x_1,\lambda)^{2n/n-2}dv_{g_0}}{\delta'(x_1,\lambda)+\delta'(x_2,\lambda)}\right\}^{2/n-2}.$$

Assume first that $d(x_1,x_2)$, the distance of x_1 and x_2, is less than $\rho/4$. Then ω_{x_1} is equal to 1 on $B(x_1,\rho/4)$ and we have:

$$(43) \quad \frac{\delta'(x_1,\lambda)}{\delta'(x_1,\lambda)+\delta'(x_2,\lambda)}(x) = \frac{u_{x_1}(x)\delta(x_1,\lambda)(x)}{u_{x_1}(x)\delta(x_1,\lambda)(x)+u_{x_2}(x)\delta(x_2,\lambda)(x)}$$
$$= \frac{\frac{u_{x_1}(x)}{u_{x_2}(x)}\delta(x_1,\lambda)(x)}{\frac{u_{x_1}(x)}{u_{x_2}(x)}\delta(x_1,\lambda)(x)+\delta(x_2,\lambda)(x)} \qquad \forall x \in B(x_1,\rho/4).$$

The continuity of u_y with respect to y implies the existence of $\eta > 0$ such that:

$$(44) \qquad \frac{1}{2} \leq \frac{u_{x_1}(x)}{u_{x_2}(x)} \leq 2 \qquad \text{if } x_2 \in B(x_1,\eta).$$

Thus if $d(x_1, x_2) \leq \eta$, we have:

$$(45) \qquad J(\sum_{i=1}^{p} \alpha_i \hat{\delta}(x_i, \lambda))$$

$$\leq \left(1 + O(\frac{1}{\lambda^{n-2}})\right) \cdot \left\{(p-1)S^{n-2/2} + \int_{B(x_1,\eta)} \frac{2\delta(x_1,\lambda) \cdot \delta(x_1,\lambda)^{2n/n-2} dv_{g_0}}{2\delta(x_1,\lambda) + \delta(x_2,\lambda)}\right\}^{2/n-2}$$

$$\leq \left(1 + O(\frac{1}{\lambda^{n-2}})\right) \cdot \left\{(p-1)S^{n-2/2} + \int_{\mathbb{R}^n} \frac{2\delta(x_1,\lambda) \cdot \delta(x_1,\lambda)^{2n/n-2} dx}{2\delta(x_1,\lambda) + \delta(x_2,\lambda)}\right\}^{2/n-2}.$$

Observe that

$$(46) \qquad \int_{\mathbb{R}^n} \frac{2\delta(x_1,\lambda) \cdot \delta(x_1,\lambda)^{2n/n-2} dx}{2\delta(x_1,\lambda) + \delta(x_2,\lambda)}$$

$$= \int_{\mathbb{R}^n} \delta(x_1,\lambda)^{2n/n-2} dx - \int_{\mathbb{R}^n} \frac{\delta(x_2,\lambda)}{2\delta(x_1,\lambda) + \delta(x_2,\lambda)} \delta(x_1,\lambda)^{2n/n-2} dx.$$

Let $\varepsilon_0 > 0$ be given and let:

$$(47) \qquad E_0 = \{\, x \in \mathbb{R}^n \text{ such that } \delta(x_1,\lambda)(x) \geq \varepsilon_0 (2\delta(x_1,\lambda) + \delta(x_2,\lambda))(x) \,\}.$$

We have:

$$(48) \qquad \int_{\mathbb{R}^n} \frac{\delta(x_2,\lambda)}{2\delta(x_1,\lambda) + \delta(x_2,\lambda)} \delta(x_1,\lambda)^{2n/n-2} dx \geq \varepsilon_0 \int_{E_0} \delta(x_1,\lambda)^{n+2/n-2} \delta(x_2,\lambda) dx$$

$$\geq \varepsilon_0 \left\{ \int_{\mathbb{R}^n} \delta(x_1,\lambda)^{n+2/n-2} \delta(x_2,\lambda) dx - \int_{\mathbb{R}^n - E_0} \delta(x_1,\lambda)^{n+2/n-2} \delta(x_2,\lambda) dx \right\}$$

$$\geq \varepsilon_0 \Big\{ \int_{\mathbb{R}^n} \delta(x_1,\lambda)^{n+2/n-2} \delta(x_2,\lambda) dx$$

$$- (\frac{\varepsilon_0}{1-2\varepsilon_0})^{2/n-2} \int_{\mathbb{R}^n} \delta(x_1,\lambda)^{n/n-2} \delta(x_2,\lambda)^{n/n-2} dx \Big\}$$

$$\geq \varepsilon_0 \Big\{ \frac{1}{2} \int_{\mathbb{R}^n} \delta(x_1,\lambda)^{n+2/n-2} \delta(x_2,\lambda) dx + \frac{1}{2} \int_{\mathbb{R}^n} \delta(x_2,\lambda)^{n+2/n-2} \delta(x_1,\lambda) dx$$

$$- (\frac{\varepsilon_0}{1-2\varepsilon_0})^{2/n-2} \int_{\mathbb{R}^n} (\, \delta(x_1,\lambda)^{n+2/n-2} \delta(x_2,\lambda) + \delta(x_2,\lambda)^{n+2/n-2} \delta(x_1,\lambda) \,) dx \Big\}$$

$$\geq \frac{\varepsilon_0}{2} \Big\{ 1 - 2(\frac{\varepsilon_0}{1-2\varepsilon_0})^{2/n-2} \Big\} \int_{\mathbb{R}^n} \Big\{ \delta(x_1,\lambda)^{n+2/n-2} \delta(x_2,\lambda)$$

$$+ \delta(x_2,\lambda)^{n+2/n-2} \delta(x_1,\lambda) \Big\} dx.$$

In (48), we used the identity

$$\int_{\mathbb{R}^n} \delta(x_1,\lambda)^{n+2/n-2}\delta(x_2,\lambda)dx = \int_{\mathbb{R}^n} \delta(x_2,\lambda)^{n+2/n-2}\delta(x_1,\lambda)dx.$$

Thus, if ε_0 is small enough, we have:

$$\begin{aligned}
(49) \qquad & \int_{\mathbb{R}^n} \frac{\delta(x_2,\lambda)}{2\delta(x_1,\lambda)+\delta(x_2,\lambda)}\delta(x_1,\lambda)^{2n/n-2}dx \\
&\geq \frac{\varepsilon_0}{4}\int_{\mathbb{R}^n} (\,\delta(x_1,\lambda)^{n+2/n-2}\delta(x_2,\lambda)+\delta(x_2,\lambda)^{n+2/n-2}\delta(x_1,\lambda)\,)dx \\
&\geq \frac{\varepsilon_0}{4}\left\{\int_{B(x_1,\rho/2)} \delta(x_1,\lambda)^{n+2/n-2}\delta(x_2,\lambda)dx + \int_{B(x_2,\rho/2)} \delta(x_2,\lambda)^{n+2/n-2}\delta(x_1,\lambda)dx\right\} \\
&\geq \bar{C}\varepsilon_0\left\{\int_{B(x_1,\rho/2)} \hat{\delta}(x_1,\lambda)^{n+2/n-2}\hat{\delta}(x_2,\lambda)dv_g + \int_{B(x_2,\rho/2)} \hat{\delta}(x_2,\lambda)^{n+2/n-2}\hat{\delta}(x_1,\lambda)dv_g\right\} \\
&\geq \bar{C}'\varepsilon_0\int_M (\,\hat{\delta}(x_1,\lambda)^{n+2/n-2}\hat{\delta}(x_2,\lambda)+\hat{\delta}(x_2,\lambda)^{n+2/n-2}\hat{\delta}(x_1,\lambda)\,)dv_g.
\end{aligned}$$

The last inequality is a straightforward estimate. By assumption, we know that

$$\begin{aligned}
&\sum_{i\neq j}\int_M \hat{\delta}(x_i,\lambda)^{n+2/n-2}\hat{\delta}(x_j,\lambda)dv_g \\
&\leq p^2\int_M (\,\hat{\delta}(x_1,\lambda)^{n+2/n-2}\hat{\delta}(x_2,\lambda)+\hat{\delta}(x_2,\lambda)^{n+2/n-2}\hat{\delta}(x_1,\lambda)\,)dv_g.
\end{aligned}$$

We derive, thus, from (45), (46), (48) and (49):

$$\begin{aligned}
(50) \qquad & J\Big(\sum_{i=1}^{p}\alpha_i\hat{\delta}(x_i,\lambda)\Big) \\
&\leq \left(1+O\Big(\frac{1}{\lambda^{n-2}}\Big)\right)\cdot\left\{pS^{n-2/2}-\frac{\bar{C}'\varepsilon_0}{p^2}\sum_{i\neq j}\int_M \hat{\delta}(x_i,\lambda)^{n+2/n-2}\hat{\delta}(x_j,\lambda)dv_g\right\}^{2/n-2} \\
&\leq p^{2/n-2}S\left(1+O\Big(\frac{1}{\lambda^{n-2}}\Big)-\frac{\bar{C}''\varepsilon_0\varepsilon_1}{p^3}\right).
\end{aligned}$$

(50) clearly implies (i), which is therefore proven if $d(x_1, x_2) \leq \eta$. If $d(x_1, x_2) \geq \eta$, then $\int_M (\hat{\delta}(x_1, \lambda)^{n+2/n-2}\hat{\delta}(x_2, \lambda) + \hat{\delta}(x_2, \lambda)^{n+2/n-2}\hat{\delta}(x_1, \lambda))dv_g$ is $O(\frac{1}{\lambda^{n-2}})$; therefore

$$\sum_{i \neq j} \int_M \hat{\delta}(x_i, \lambda)^{n+2/n-2}\hat{\delta}(x_j, \lambda)dv_g \leq p^2 \cdot O(\frac{1}{\lambda^{n-2}}) = O'(\frac{1}{\lambda^{n-2}}).$$

Taking λ large enough, we have $\sum_{i \neq j} \int_M \hat{\delta}(x_i, \lambda)^{n+2/n-2}\hat{\delta}(x_j, \lambda)dv_g \leq O'(\frac{1}{\lambda^{n-2}}) < \varepsilon_1$, and therefore the proof of (i) reduces to the case $d(x_1, x_2) \leq \eta$. (i) is thereby established. The proof of (ii) and the proof of (iii) will be completed together since they both rest on an expansion of J. First, we recall the following result from [**7**], the proof of which is provided in Appendix B:

Lemma 6. *Let $q > 2$ be given. There exists $\gamma > 1$ such that for any a_1, ..., $a_p \geq 0$, we have:*

$$(\Sigma_{i=1}^p a_i)^q \geq \Sigma_{i=1}^p a_i^q + \frac{\gamma q}{2} \Sigma_{i \neq j} a_i^{q-1} a_j.$$

Using Lemma 6, we derive the following inequality:

$$J(\sum_{i=1}^p \alpha_i \hat{\delta}(x_i, \lambda)) \leq \tag{51}$$

$$\frac{\left(\sum_{i=1}^p \alpha_i^2 \int_M -L\hat{\delta}(x_i, \lambda)\hat{\delta}(x_i, \lambda)dv_g + 2\sum_{i<j} \alpha_i\alpha_j \int_M -L\hat{\delta}(x_i, \lambda)\hat{\delta}(x_j, \lambda)dv_g\right)^{n/n-2}}{\sum_{i=1}^p \alpha_i^{2n/n-2} \int_M \hat{\delta}(x_i, \lambda)^{2n/n-2}dv_g + \sum_{i \neq j}^p \frac{\gamma n \alpha_i^{n+2/n-2}\alpha_j}{n-2} \int_M \hat{\delta}(x_i, \lambda)^{n+2/n-2}\hat{\delta}(x_j, \lambda)dv_g}.$$

(51) implies using Lemmas 3 and 4:

$$J(\sum_{i=1}^p \alpha_i \hat{\delta}(x_i, \lambda)) \leq \tag{52}$$

$$\frac{\left(\sum_{i=1}^p \alpha_i^2 S^{n-2/2} + 2(1+\theta)\sum_{i<j} \alpha_i\alpha_j \int_M \hat{\delta}(x_i, \lambda)^{n+2/n-2}\hat{\delta}(x_j, \lambda)dv_g + \frac{\bar{c}'\Sigma\alpha_i^2}{\lambda^{n-2}}\right)^{n/n-2}}{\sum \alpha_i^{2n/n-2} S^{n-2/2} + \sum_{i \neq j} \frac{\gamma n \alpha_i^{n+2/n-2}\alpha_j}{n-2} \int_M \hat{\delta}(x_i, \lambda)^{n+2/n-2}\hat{\delta}(x_j, \lambda)dv_g - \frac{\bar{c}'\Sigma\alpha_i^{2n/n-2}}{\lambda^{n-2}}},$$

where $\bar{c}'$ is a constant independent of $\alpha_1, \ldots, \alpha_p$, λ and p. Let us assume that

$$(53) \qquad \sum_{i \neq j} \int_M \hat{\delta}(x_i,\lambda)^{n+2/n-2}\hat{\delta}(x_j,\lambda)dv_g < \varepsilon_1.$$

Let $1 < \gamma' < \gamma$. If ε_1 is chosen small enough, then, for λ large enough, we have:

$$(54) \qquad J(\sum_{i=1}^{p} \alpha_i \hat{\delta}(x_i,\lambda)) \leq$$

$$\frac{(\sum_{i=1}^p \alpha_i^2)^{n/n-2}}{\sum_{i=1}^p \alpha_i^{2n/n-2}} S\Big\{1 + \frac{2n}{n-2}(1+\theta)\frac{\sum_{i<j}\alpha_i\alpha_j \int_M \hat{\delta}(x_i,\lambda)^{n+2/n-2}\hat{\delta}(x_j,\lambda)dv_g}{\sum \alpha_r^2 S^{n-2/2}}$$

$$- \frac{\gamma' n}{n-2}\cdot\frac{\sum_{i\neq j}\alpha_i^{n+2/n-2}\alpha_j}{S^{n-2/2}\sum_{i=1}^p \alpha_i^{2n/n-2}}\int_M \hat{\delta}(x_i,\lambda)^{n+2/n-2}\hat{\delta}(x_j,\lambda)dv_g + \frac{3\bar{c}'}{\lambda^{n-2}}\Big\}.$$

(iii) follows easily from (54). Assume now that we choose $0 < \theta_0 < 1$ and $0 < \theta$ so that

$$(55) \qquad \frac{n}{n-2}\cdot\frac{1+\theta}{\theta_0^2} < \frac{\gamma' n \theta_0^{2n/n-2}}{n-2}.$$

Then, assuming $\alpha_i/\alpha_j \geq \theta_0$ for any $i \neq j$ and assuming λ is large enough, we have, with $\delta = \frac{\gamma' n}{n-2}\cdot\theta_0^{2n/n-2} - \frac{n}{n-2}\frac{1+\theta}{\theta_0^2}$:

$$(56) \qquad J(\sum_{i=1}^{p}\alpha_i\hat{\delta}(x_i,\lambda))$$

$$\leq \frac{(\sum_{i=1}^p \alpha_i^2)^{n/n-2}}{\sum_{i\neq j}\alpha_i^{2n/n-2}} S\Big\{1 + \frac{3\bar{c}'}{\lambda^{n-2}} - \frac{\delta}{p}\sum_{i\neq j}\int_M \hat{\delta}(x_i,\lambda)^{n+2/n-2}\hat{\delta}(x_j,\lambda)dv_g\Big\}.$$

Using then (v) of Lemma 4, we derive (ii) of Lemma 5. The proof of Lemma 5 is thereby complete.

Appendix A

We want to prove that

$$\text{(A1)} \qquad \frac{1}{\lambda^{n-2/2}} \int_{B(x_0,\rho)} \delta(x_0,\lambda)^{4/n-2} \delta(x_1,\lambda) dv_{g_0} = o\Big(\int_{\mathbf{R}^n} \delta(x_0,\lambda)^{n+2/n-2} \delta(x_1,\lambda) dv_{g_0} \Big).$$

We pick $\varepsilon > 0$, a fixed number. Let

$$\text{(A2)} \qquad A_\varepsilon = \Big\{ x \in B(x_0,\rho)\,;\, \delta(x_0,\lambda)(x) \geq (\varepsilon \lambda^{n-2/2})^{-1} \Big\}.$$

We then have:

$$\text{(A3)} \qquad \frac{1}{\lambda^{n-2/2}} \int_{B(x_0,\rho)} \delta(x_0,\lambda)^{4/n-2} \delta(x_1,\lambda) dv_{g_0}$$

$$\leq \varepsilon \int_{A_\varepsilon} \delta(x_0,\lambda)^{n+2/n-2} \delta(x_1,\lambda) dv_{g_0} + \int_{A_\varepsilon^c} \frac{\delta(x_0,\lambda)^{4/n-2} \delta(x_1,\lambda) dv_{g_0}}{\lambda^{n-2/2}}$$

$$\leq \varepsilon \int_{A_\varepsilon} \delta(x_0,\lambda)^{n+2/n-2} \delta(x_1,\lambda) dv_{g_0} + \int_{B(x_0,\rho)} \frac{\delta(x_1,\lambda) dv_{g_0}}{\lambda^2 \lambda^{n-2/2} \varepsilon^{4/n-2}}$$

$$\leq \varepsilon \int_{\mathbf{R}^n} \delta(x_0,\lambda)^{n+2/n-2} \delta(x_1,\lambda) dv_{g_0} + \frac{C(\rho) \Big(\int_{\mathbf{R}^n} \delta(x_1,\lambda)^{n+2/n-2} dv_{g_0} \Big)^{n/n-2}}{\lambda^2 \lambda^{n-2/2} \varepsilon^{4/n-2}}$$

$$\leq \varepsilon \int_{\mathbf{R}^n} \delta(x_0,\lambda)^{n+2/n-2} \delta(x_1,\lambda) dv_{g_0} + \frac{C'(\rho)}{\varepsilon^{4/n-2} \lambda^{n+2/2} \lambda^{(n-2)^2/2(n+2)}}.$$

Since $\frac{n+2}{2} + \frac{(n-2)^2}{2(n+2)} > n-2$ and since

$$\int_{\mathbf{R}^n} \delta(x_0,\lambda)^{n+2/n-2} \delta(x_1,\lambda) dv_{g_0} \geq \frac{C}{\lambda^{n-2}}, \quad C \text{ a universal positive constant,}$$

(A3) implies (A1); indeed, for any $\eta > 0$, it suffices first to pick an appropriate ε small and then choose λ large so that (A3) implies:

$$\text{(A4)} \qquad \frac{1}{\lambda^{n-2/2}} \int_{B(x_0,\rho)} \delta(x_0,\lambda)^{4/n-2} \delta(x_1,\lambda) dv_g \leq \eta \int_{\mathbf{R}^n} \delta(x_0,\lambda)^{n+2/n-2} \delta(x_1,\lambda) dv_g.$$

This exactly means that

$$\frac{1}{\lambda^{n-2/2}} \int_{B(x_0,\rho)} \delta(x_0,\lambda)^{4/n-2} \delta(x_1,\lambda) dv_{g_0} = o\Big(\int_{\mathbf{R}^n} \delta(x_0,\lambda)^{n+2/n-2} \delta(x_1,\lambda) dv_{g_0} \Big). \quad \blacksquare$$

Appendix B

Proof of Lemma 6:

We take $p = 2$. The general case follows by induction (γ remains unchanged), using the fact that $q - 1 \geq 1$ and

$$(a+b)^{q-1} \geq a^{q-1} + b^{q-1} \quad (a, b \geq 0).$$

We thus want to prove:

$$(a+b)^q - a^q - b^q \geq \frac{\gamma q}{2}(a^{q-1}b + b^{q-1}a).$$

The function $t \mapsto (t+1)^q - t^q - 1 = \varphi(t)$ is convex ($q > 2$). We thus have:

$$\varphi(t) - \varphi(0) \geq \varphi'(0)t = q\,t$$

hence

$$(t+1)^q - t^q - 1 \geq q\,t.$$

The strict convexity of φ for $t > 0$ implies in fact that:

$$(t+1)^q - t^q - 1 > q\,t \quad \text{for } t > 0.$$

Changing $t \to 1/t$, we derive:

$$(t+1)^q - t^q - 1 > q\,t^{q-1}.$$

Adding these two inequalities, we derive:

$$(t+1)^q - t^q - 1 > \frac{q}{2}(t + t^{q-1})$$

which provides on each compact interval of $\mathbb{R}^+ - \{0\}$ a constant γ such that the inequality of Lemma 6 holds. We are thus led to compute

$$\lim_{t \to 0 \text{ or } t \to \infty} \frac{(t+1)^q - t^q - 1}{t + t^{q-1}}.$$

This limit is easily seen (change t in $1/t$ if $t \to \infty$) to be equal to $q > q/2$. Lemma 6 follows.

Appendix C

Assuming that (M^n, g) is locally conformally flat, we want to prove that we may find a euclidian metric $\bar{g}_0$, conformal to g on a neighborhood of a point x_0 of M, depending smoothly on x_0. Let x_0 be a point of M, $\bar{g}$ be a euclidian metric conformal to g in a neighborhood of x_0 and $\bar{g}' = e^f \bar{g}$ another such euclidian metric. Let x_i, $i = 1, \ldots, n$, be orthonormal coordinates for $\bar{g}$, around x_0. Denoting $(\bar{g}_{ij})$ the metric tensor in the basis $(\partial/\partial x^1, \ldots, \partial/\partial x^n)$, we have:

$$\bar{g}_{ij} = \delta_i^j \tag{C1}$$

where δ_i^j is the Kronecker symbol of i and j ($\delta_i^j = 1$ if $i = j$; $\delta_i^j = 0$ otherwise). If φ is a function defined on a neighborhood of x_0, we denote

$$\partial_i \varphi = D\varphi\,(\partial/\partial x^i) \tag{C2}$$

where $D\varphi$ is the differential of φ. Let $\bar{\Gamma}_{ij}^{\ell}$ denote the Christoffel symbols of $\bar{g}$ and $\bar{\Gamma}'^{\ell}_{ij}$ be those of $\bar{g}'$. Since (x^i) is a euclidian system of coordinates for $\bar{g}$, we have:

$$\bar{\Gamma}_{ij}^{\ell} = 0. \tag{C3}$$

The formula for $\bar{\Gamma}'^{\ell}_{ij}$ is (see [8] page 125):

$$\bar{\Gamma}'^{\ell}_{ij} = \bar{\Gamma}_{ij}^{\ell} + \frac{1}{2}(\delta_i^{\ell}\partial_i f + \delta_i^{\ell}\partial_j f - g_{ij}\partial_{\ell} f) = \frac{1}{2}(\delta_j^{\ell}\partial_i f + \delta_i^{\ell}\partial_j f - \delta_i^j\partial_{\ell} f). \tag{C4}$$

Since $\bar{g}'$ is euclidian, the curvature tensor R'^{ℓ}_{kij} of $\bar{g}'$ vanishes and this is equivalent to the fact that $\bar{g}'$ is euclidian. The formula for R'^{ℓ}_{kij} is (see [8] page 4)

$$R'^{\ell}_{kij} = \partial_i \Gamma'^{\ell}_{jk} - \partial_j \Gamma'^{\ell}_{ik} + \Gamma'^{\ell}_{im}\Gamma'^{m}_{jk} - \Gamma'^{\ell}_{jm}\Gamma'^{m}_{ik}. \tag{C5}$$

Hence, using (C4), we have:

$$R'^{\ell}_{kij} = \frac{1}{2}\partial_i(\delta^\ell_k\partial_j f + \delta^\ell_j\partial_k f - \delta^k_j\partial_\ell f) - \frac{1}{2}\partial_j(\delta^\ell_k\partial_i f + \delta^\ell_i\partial_k f - \delta^k_i\partial_\ell f)$$
$$+ \frac{1}{4}(\delta^\ell_m\partial_i f + \delta^\ell_i\partial_m f - \delta^m_i\partial_\ell f)(\delta^m_k\partial_j f + \delta^m_j\partial_k f - \delta^k_j\partial_m f)$$
$$- \frac{1}{4}(\delta^\ell_m\partial_j f + \delta^\ell_j\partial_m f - \delta^m_j\partial_\ell f)(\delta^m_k\partial_i f + \delta^m_i\partial_k f - \delta^k_i\partial_m f). \tag{C6}$$

Using (C5), we easily derive:

$$R'^{\ell}_{kji} = -R'^{\ell}_{kij}. \tag{C7}$$

Therefore, $R'^{\ell}_{kii} = 0$ and i and j may be interchanged in the arguments below, involving the vanishing of the curvature tensor. From (C6), it is also easy to see that R'^{ℓ}_{kji} is zero if

$$i \neq j, \ \ k \neq i, \ \ k \neq j; \ \ \ell \neq i; \ \ \ell \neq j. \tag{C8}$$

We are thus left with two cases, the other cases being derived by interchanging i and j and using (C7):

(Case 1) $\quad i \neq j, \ \ k = i.$

(Case 2) $\quad i \neq j, \ \ k \neq i, \ \ k \neq j, \ \ \ell = i.$

In Case 1, we have:

$$R'^{\ell}_{iij} = \frac{1}{2}\delta^\ell_j\partial_i\partial_k f - \frac{1}{2}\delta^\ell_i\partial_j\partial_k f + \frac{1}{2}\partial_j\partial_\ell f$$
$$+ \frac{1}{4}\delta^\ell_i\partial_j f\partial_k f - \frac{1}{4}\delta^\ell_j\partial_i f\partial_k f + \frac{1}{4}\delta^\ell_j|\partial_m f|^2 - \frac{1}{4}\partial_j f\partial_\ell f$$
$$= \{ -\partial_j\partial_\ell(e^{-f/2}) + \delta^\ell_i\partial_j\partial_k(e^{-f/2}) - \delta^\ell_j\partial_i\partial_k(e^{-f/2}) \} e^{f/2}$$
$$+ \frac{1}{4}\delta^\ell_j|\partial_m f|^2 = 0. \tag{C9}$$

We subdivide the argument in three subcases:

subcase a: if we assume that $\ell \neq j$ and $\ell \neq i$, then $\delta_i^\ell = \delta_j^\ell = 0$ and (C9) becomes:

$$\partial_j \partial_\ell (e^{-f/2}) = 0 \tag{C10}$$

where i, j, ℓ are pairwise distinct. Since $n \geq 3$, this does not imply any restriction and (C10) is equivalent to:

$$\partial_i \partial_j (e^{-f/2}) = 0, \qquad \forall i \neq j. \tag{C11}$$

Hence,

$$e^{-f/2} = f_1(x^1) + \cdots + f_n(x^n) \tag{C12}$$

where the f_i's are smooth functions of one real variable.

subcase b: if we assume that $\ell = i$, we immediately see from (C9) that

$$R'^{i}_{iij} = 0.$$

subcase c: lastly, if we assume $\ell = j$ ($\ell \neq i$), we then derive from (C9):

$$\{-\partial_i^2(e^{-f/2}) - \partial_j^2(e^{-f/2})\}e^{f/2} + \frac{1}{4}|\partial_m f|^2 = 0. \tag{C13}$$

(C13) implies, since (i, j) is an arbitrary couple of distinct indexes:

$$\partial_i^2(e^{-f/2}) = \partial_j^2(e^{-f/2}), \qquad \forall i \neq j. \tag{C14}$$

Using (C12), we derive:

(C15) $$f_i''(x^i) = f_j''(x^j) = C.$$

Thus:

(C16) $$e^{-f/2} = \Sigma(\frac{C(x^i)^2}{2} + D_i x^i + E_i).$$

Using (C16), (C13) reads:

(C17) $$2C\Sigma(\frac{C(x^i)^2}{2} + D_i x^i + E_i) = \Sigma(Cx^i + D_i)^2.$$

Hence:

(C18) $$2C(\Sigma E_i) = \Sigma D_i^2.$$

Case 1 thus implies that:

(C19) $$\begin{cases} e^{-f/2} = \Sigma(\frac{C(x^i)^2}{2} + D_i x^i + E_i) \\ 2C(\Sigma E_i) = \Sigma D_i^2. \end{cases}$$

In Case 2, we have:

(C20) $$R'^{i}_{kij} = -\frac{1}{2}\partial_j\partial_k f + \frac{1}{4}\partial_k f \partial_j f = \partial_j\partial_k(e^{-f/2})e^{f/2} = 0.$$

Since k and j are distinct, (C19) implies (C20). Thus, we have established that a necessary and sufficient condition for $\bar{g}' = e^f \bar{g}$ to be euclidian is that $e^{-f/2}$ satisfies (C19). Let then $\gamma_{\bar{g}'}(x)$ be the function:

(C21) $$\gamma_{\bar{g}'}(x) = dv_{\bar{g}'}(x)/dv_g(x).$$

We define $\bar{g}_0$, in a neighborhood of x_0, as the unique euclidian metric conformal to g such that:

$$\text{(C22)} \qquad \gamma_{\bar{g}_0}(x_0) = 1; \; D_{x_0}\gamma_{\bar{g}_0} = 0,$$

where $D_{x_0}\gamma_{\bar{g}_0}$ is the differential of $\gamma_{\bar{g}_0}$ at x_0. Assuming that $\gamma_{\bar{g}}(x_0) = 1$ and setting $\bar{g}_0 = e^f g$, where $e^{-f/2}$ satisfies (C19), the above (C22) is equivalent to:

$$\text{(C23)} \qquad \Sigma E_i = 1; \; D_{x_0}e^f + e^f(x_0)D_{x_0}\gamma_{\bar{g}} = 0,$$

hence

$$\text{(C24)} \qquad \Sigma E_i = 1; \; 2D_i = D_{x_0}\gamma_{\bar{g}}(\partial/\partial x^i).$$

Thus, the value of ΣE_i and of D_i is completely determined and $\bar{g}_0$ is therefore unique and depends smoothly on x_0. ∎

References

[1] T. Aubin, *Equations differentielles non linéaires et problèmes de Yamabe concernant la courbure scalaire*, J. Math. Pures et Appl. **55** (1976), 269–296.

[2] R. Schoen, *Conformal deformation of a metric to constant scalar curvature*, J. Diff. Geom. **20** (1984), 479–495.

[3] R. Schoen and S. T. Yau, *Conformally flat manifolds, Kleinnian groups and scalar curvature*, Invent. Math. **92** (1988), 47–71.

[4] A. Bahri, "Pseudo-Orbits of contact forms," Pitman Research Nothes in Math. **173**, Longman Scientific and Technical.

[5] A. Bahri, "Critical points at infinity in some variational problems," Pitman Research Nothes in Math. **182**, Longman Scientific and Technical.

[6] A. Bahri and J. M. Coron, *On a nonlinear elliptic equation involving the limiting Sobolev exponent*, Comm. Pure Appl. Math. **41** (1988), 253–294.

[7] A. Bahri and H. Brezis, *Equations elliptiques non linéaires sur des variétés avec exposant limite de Sobolev*, C. R. Acad. Sc., Ser. I, **307** (1988), 573–576.

[8] T. Aubin, "Nonlinear Analysis on Manifolds," Grundlehren der mathematischen Wissenschaften **252**, Springer-Verlag.

2

Einstein-Hermitian Metrics on Non-Compact Kähler Manifolds

Shigetoshi Bando
Mathematical Institute
Tohoku University
Sendai, 980, Japan

In his paper [D2], Donaldson showed that there is a natural correspondence between the moduli space of anti-self-dual connections with finite action on the complex Euclidean plane $\mathbf{C}^2$ and the moduli space of holomorphic vector bundles on the complex projective plane $\mathbf{P}^2$ whose restrictions to the complex line at infinity are trivial. The purpose of the paper is to generalize the result for certain class of Kähler manifolds.

Let $\bar{X}$ be an n-dimensional ($n \geq 2$) compact Kähler manifold and D a smooth divisor which has positive normal line bundle. We denote the complement of D by X and put a cone-like Kähler metric ω on X. We fix a point o in X and denote the distance from o by r. Then our main result is

Theorem 1. *There is a natural correspondence between the moduli space of Einstein-Hermitian holomorphic vector bundles on* (X, ω) *which satisfy the curvature decay condition*

$$|F| = O(r^{-2-\epsilon}), \qquad \text{with } \epsilon > 0,$$

and have trivial holonomy at infinity, and the moduli space of holomorphic vector bundles on $\bar{X}$ *whose restrictions to* D *are* $U(r)$*-flat.*

Corollary 2. *If* (X, ω) *is asymptotically locally Euclidean,* ALE *in short, then in the Theorem 1 we can replace the curvature decay condition by*

$$\int_X |F|^n < \infty.$$

And in this case it is equivalent to

$$|F| = O(r^{-(2n-\epsilon)}), \qquad \text{for any } \epsilon > 0.$$

Corollary 3. *Let* X *be a non-singular compact Kähler surface,* C *a non-singular curve with positive self intersection* $C^2 > 0$ *and* E *a holomorphic vector bundle on* X*. If the restriction* $E|_C$ *of* E *is poly-stable with vanishing first Chern class* $c_1(E|_C) = 0 \in H^2(C, \mathbf{R})$ *then we have*

$$2c_2(E) - c_1(E)^2 \geq 0.$$

Moreover the equality holds if and only if E *is flat.*

Remark.

(i) In Corollary 3, if $E|_C$ is poly-stable but may have non-vanishing first Chern class, then considering $E \otimes E^*$ one can get the following inequality.

$$2rc_2(E) - (r-1)c_1(E)^2 \geq 0, \qquad r = \text{rank} E.$$

One can also show that the equality holds if and only if E is projectively flat.

(ii) Let X is a compact normal surface, C a smooth ample divisor and E a holomorphic vector bundle on X whose restriction to C is poly-stable with first Chern class zero. If we take a resolution, we can apply Corollary 3.

Here we remark that Theorem 1 can be considered as a sort of removable singularity theorem of holomorphic vector bundles across divisors. For a removable singularity theorem across subvarieties of higher co-dimension, the readers are referred to [B] and [B-S].

The author would like to express his thanks to Professors Mabuchi and Nakajima for helpful discussions. He also would like to acknowledge his gratitude to Max-Planck-Institut für Mathematik for hospitality. This work was done during his stay in Bonn.

1. Existence of Einstein-Hermitian metrics

Let $(\bar{X},\omega_0)$ be a compact n-dimensional ($n \geq 2$) Kähler manifold and D a smooth divisor which has positive normal line bundle. We denote the Poincaré dual of D and its restriction to D by the same notation $[D]$. Set $X = \bar{X}\backslash D$. By assumption, there exists a Hermitian metric $\|\cdot\|$ on the line bundle L_D defined by D such that its curvature form θ is positive definite on a neighborhood of D. Pick a holomorphic section σ of L_D on $\bar{X}$ whose zero divisor is D. Put $t = \log\|\sigma\|^{-2}$. Then $\theta = \sqrt{-1}\partial\bar{\partial}\log\|\sigma\|^{-2} = \sqrt{-1}\partial\bar{\partial}t$. Fix an arbitrary positive number $a > 0$ and a sufficiently large positive constant C. We define a Kähler metric ω on X by

$$\begin{aligned}\omega &= \sqrt{-1}\partial\bar{\partial}\frac{1}{a}\exp(at) + C\omega_0\\ &= \exp(at)\,\theta + a\exp(at)\sqrt{-1}\partial t \wedge \bar{\partial}t + C\omega_0.\end{aligned}$$

Here we identify a Kähler metric and its Kähler form. Then it is easy to see that ω is a $C^{k,\alpha}$-cone-like metric for any positive integer k, real number $0 < \alpha < 1$ and some positive number $\tau > 0$.

Definition. A complete Riemannian manifold (M,g) is said to be $C^{k,\alpha}$-cone-like of order $\tau > 0$ if there exists a compact subset K of M, a compact Riemannian manifold (N,h), a compact subset K' of the cone CN over N and a diffeomorphism $\phi : CN\backslash K' \longrightarrow M\backslash K$ such that it holds that up to $C^{k,\alpha}$-order

$$\phi^* g = dr^2 + r^2 h + O(r^{-\tau}).$$

Then in particular (M,g) is asymptotically flat in the following sense ([BK]).

Definition. A complete Riemannian metric g on a manifold M is said to be of $C^{k,\alpha}$-asymptotically flat geometry if for each point $p \in M$ with distance r from a fixed point o in M, there exists a harmonic coordinates system $x = (x^1, x^2, \ldots, x^m)$ centered at p which satisfies the following conditions:

(i) The coordinate x runs over a unit ball $\mathbf{B}^m$ in $\mathbf{R}^m$.

(ii) If we write $g = \sum g_{ij}(x)dx^i dx^j$, then the matrix $(r^2+1)^{-1}(g_{ij})$ is bounded from below by a constant positive matrix independent of p.
(ii) The $C^{k,\alpha}$-norms of $(r^2+1)^{-1}g_{ij}$, as functions in x, are uniformly bounded.

On such a manifold we can define the Banach space $C^{k,\alpha}_\delta$ of weighted $C^{k,\alpha}$-bounded functions: We may assume that $(r^2+1)^{-1}(g_{ij}) \leq 1/2(\delta_{ij})$ and the norm of a function $u \in C^{k,\alpha}_\delta$ is given by the supremum of the $C^{k,\alpha}$-norms of $(r^2+1)^{\delta/2}u$ with respect to the coordinates x. Then we can apply the interior Schauder estimates as in [BK]. Note that on a cone-like manifold the Sobolev inequality holds.

Definition. For the Hermitian holomorphic vector bundle (E, h) on (X, ω) with fast decreasing curvature $|F| = O(r^{-2-\epsilon})$, $\epsilon > 0$, we can define a holonomy at infinity as follows. Take a complex disk Δ in $\bar{X}$ which transversally intersects D at a point o and the circle $S_r \subset \Delta$ of radius r centered at o. Then the equivalence class of the holonomy of S_r converges as r tends to zero. It is easy to see that the equivalence class is independent of the choice of the disk, and we call it the holonomy at infinity.

Theorem 4. *Let E a holomorphic vector bundle on $\bar{X}$. If its restriction $E|_D$ to D is poly-stable and degree zero with respect to $[D]$, then the restriction $E|_X$ admits an Einstein-Hermitian metric with respect to the metric ω which satisfies the curvature decay condition*

$$|F| = O(r^{-2}).$$

Moreover, if $E|_D$ is flat, then the Einstein-Hermitian metric satisfies

$$|F| = O(r^{-2-\epsilon}),$$

and has trivial holonomy at infinity.

Proof. By assumption, $E|_D$ admits an Einstein-Hermitian metric h_0 with respect to the Kähler metric $\theta|_D$ (cf. [N-S], [D1, 3–4], [U-Y1–2], [Sim], [Siu], [K]). We smoothly extend it over $\bar{X}$ and get a Hermitian metric, we still call it h_0, on E. Then it is easy to see that with respect to the metric ω, its curvature F_0 satisfies with some $0 < \epsilon < 1$

$$|F_0| = O(r^{-2}), \qquad (|F_0| = O(r^{-2-\epsilon}), \quad \text{if } E|_D \text{ is flat}),$$
$$|\Lambda F_0| = O(r^{-2-\epsilon}),$$

together with the corresponding estimates for their covariant derivatives. Now we solve the following heat equation on Hermitian metric h.

$$\frac{dh}{dt}h^{-1} = -\sqrt{-1}\Lambda F,$$
$$h|_{t=0} = h_0.$$

As shown by Simpson [Sim], for any compact smooth subdomain K in X, we have a unique solution until infinite time with the boundary condition

$$h|_{\partial K} = h_0|_{\partial K}.$$

Then it satisfies

$$\frac{d\Lambda F}{dt} = \Box \Lambda F,$$
$$\frac{d|\Lambda F|}{dt} \leq \Box |\Lambda F|,$$
$$|\Lambda F|(t,x) \leq \int_K H_K(t,x,y)|\Lambda F_0|(y).$$

Here $\Box$ and $H_K(t,x,y)$ are the complex (crude) Laplacian with respect to ω and its heat kernel on K with the Dirichlet boundary condition. Since the metric ω is cone-like, X admits the heat kernel $H(t,x,y)$ and the Green function $G(x,y) = \int_0^\infty H(t,x,y)dt$. Then

$$\begin{aligned}\int_0^\infty |\frac{dh}{dt}h^{-1}|(x)dt &\leq \int_0^\infty dt \int_K H_K(t,x,y)|\Lambda F_0|(y)\\ &\leq \int_0^\infty dt \int_X H(t,x,y)|\Lambda F_0|(y)\\ &= \int_X G(x,y)|\Lambda F_0|(y).\end{aligned}$$

Applying the argument of [B-K] to the function $u(x) = \int_X G(x,y)|\Lambda F_0|(y)$ which satisfies $\Box u = -|\Lambda F_0|(y) = O(r^{-2-\epsilon})$, we can show the estimate $u = O(r^{-\epsilon})$. Thus taking the limit of $t \longrightarrow \infty$ and $K \longrightarrow X$, the solution metrics conveges to an Einstein-Hermitian metric. Now we call it h and it holds that $|h - h_0| = O(r^{-\epsilon})$. Then the argument of [B-K] and the proof of Proposition 1 in [B-S] gives the higher order estimates of $h - h_0$ and hence the desired curvature estimates. The triviality of the holonomy of h at infinity follows from that for h_0 and the estimate of $h - h_0$.

If $n = 2$ and $E|_D$ is flat, then it holds

$$2c_2(E) - c_1(E)^2 = (8\pi^2)^{-1} \int_X (|F|^2 - |\Lambda F|^2).$$

Since $\Lambda F = 0$, we get

$$2c_2(E) - c_1(E)^2 = (8\pi^2)^{-1} \int_X |F|^2 \geq 0.$$

This shows the first part of Corollary 3, and that the equality implies the flatness of E on X. To show the flatness on $\bar{X}$ we need results in the next section.

Remark. By the similar argument we can show the following existence theorem for harmonic mappings. Let M be a not necessarily complete Riemannian manifold with the Green function $G(x,y) \geq 0$ and N a complete Riemannian manifold with non-positive sectional curvature. We denote the distance function on N by d. For a mapping $f : M \longrightarrow N$, we define

$$u_f(x) = \int_M G(x,y)|\triangle f|(y).$$

Theorem. *If the integral u_f converges and defines a continuous function on M, then f can be deformed by the heat equation to a harmonic mapping h which satisfies*

$$d(h(x), f(x)) \leq u_f(x).$$

2. Einstein-Hermitian bundles with fast curvature decay

By Theorem 4, we get the correspondence stated in Theorem 1 in one direction. Here we work in the converse direction. Let (E, h) be an Einstein-Hermitian holomorphic vector bundle on (X, ω) whose curvature F decreases rapidly such that with $0 < \epsilon < 1$

$$|F| = O(r^{-2-\epsilon}).$$

F satisfies the following equation.

$$(*) \qquad \Box F = F * R + F * F,$$

where R is the curvature tensor of the metric ω and $*$'s stand for some bilinear pairings.

The following Lemma 5 is standerd. For instance, refer to [B-K-N].

Lemma 5. *Let u,f and g be non-negative functions and τ a constant such that*

$$\Box u \geq -fu - g, \qquad f = O(r^{-2}), \quad g = O(r^{-2-\tau}),$$
$$\frac{1}{r^{2n}} \int_{B(x,\delta r)} u^2 = O(r^{-\tau}),$$

where $B(x, \delta r)$ is the ball of radius δr centered at x with some fixed number $0 < \delta < 1$. Then u satisfies

$$u = O(r^{-\tau}).$$

Lemma 6. *For any non-negative integar k, we have*

$$|\nabla^k F| = O(r^{-2-k-\epsilon}).$$

Proof. We only show the case $k = 1$. The general case is done by induction. The equation $(*)$ implies

$$(**) \qquad \Box |F|^2 \geq |\nabla F|^2 - C(|R||F|^2 + |F|^3).$$

Here and hereafter C stands for a general constant which may change in different appearance. Fix a small $0 < \delta < 1$ and take a cut-off function $\phi \geq 0$ such that $\phi = 1$ on $B(x, \delta r)$, $d(\text{supp}\phi, o) \geq \delta r$ and $\Box \phi \leq C r^{-2}$. Multiply the inequality $(**)$ by ϕ and integrate the result by parts, then we get

$$\int_{B(x,\delta r)} |\nabla F|^2 \leq \int \Box\phi |F|^2 + C r^{-2} \int \phi |F|^2 \leq C r^{2n-6-2\epsilon}.$$

Differentiate the equation $(*)$ and get

$$\Box\nabla F = R * \nabla F + F * \nabla F + \nabla R * F$$
$$\Box|\nabla F| \geq -C(|R| + |F|)|\nabla F| - C|\nabla R||F|.$$

Then we apply Lemma 5 and conclude that

$$|\nabla F| = O(r^{-3-\epsilon}).$$

From now on we work locally. We take a local coordinates system $(z^1, z^2, \dots, z^{n-1}, z^n) = (z', z)$ at an arbitrary fixed point $p \in D$ such that $D = \{z^n = 0\}$. By calculation one can show the following Lemmas.

Lemma 7. *With respect to the flat metric $|dz'|^2 + |z^n|^{-2}|dz^n|^2$, the curvature F admits the following estimates. For any non-zero integer k*

$$|\nabla^k F| = O(|z|^{a\epsilon}).$$

We take an m-covering $\phi_m : (w', w^n) \longrightarrow (z', z^n)$ such that $z' = w'$ and $z^n = (w^n)^m$ with large positve integer m.

Lemma 8. *We pull back the bundle (E, h) to w-space by ϕ_m, then with respect to the flat metric $|dw|^2$*

$$|\nabla^k F| = O(|w|^{a\epsilon m-k-2}).$$

Now we put the assumption of trivial holonomy at infinity. On the w-space we have C^l-bound on the curvature tensor for any fixed l taking m large, the connection extends over the set $D_m = \{w^n = 0\}$ smoothly up to C^l-order (cf. [BKN]). Since outside D_m, the Hermitian connection satisfies the integrability condition, it remains so over D_m and defines a Hermitian holomorphic vector bundle E_m on the w-space. The deck transformation group $G_m = \{\tilde{\rho} : (w', w^n) \longrightarrow (w', \rho w^n) \mid \rho^m = 1\}$ lifts to a group of holomorphic bundle maps of E_m. We recover the original bundle E as the invariant subspace $E = E_m^{G_m}|_{\{w^n \neq 0\}}$. Since by assumption the isotropy group of G_m at D_m is trivial, the natural extension $\hat{E} = E_m^{G_m}$ of E over D is again a Hermitian holomorphic vector bundle. Note that $\hat{E}|_D$ and $E_m|_{D_m}$ is isomorphic and the later has vanishing curvature. Hence $\hat{E}|_D$ is a flat bundle. This completes the proof of the converse direction of Theorem 1. The proof also shows the last part of Corollary 3.

Corollary 2 follows from the results in [B-K-N, §4].

Reference

[B] S. Bando, *Removable singularities of holomorphic vector bundles*, a preprint.

[B-K] S. Bando and R. Kobayashi, *Ricci-flat Kähler metrics on affine algebraic manifolds. II*, Math. Ann. 287 (1990), 175–180.

[B-K-N] S. Bando, A. Kause and H. Nakajima, *On a construction of coordinates at infinity on manifolds with fast curvature decay and maximal volume growth*, Invent. math. 97 (1989), 313–349.

[B-S] S. Bando and Y.-T. Siu, *Stable sheaves and Einstein-Hermitian metrics*, in preparation.

[C-G] M. Cornalba and P. Griffiths, *Analytic cycles and vector bundles on non-compact algebraic varieties*, Invent. math. 28 (1975), 1–106.

[D1] S. K. Donaldson, *A new proof of a theorem of Narasimhan and Seshadri*, J. Diff. Geom. 18 (1983), 269–277.

[D2] S. K. Donaldson, *Instanton and geometric invariant theory*, Comm. Math. Phys. 93 (1984), 453–460.

[D3] S. K. Donaldson, *Anti self-dual Yang-Mills connections over comlpex algebraic surfaces and stable vector bundles*, Proc. London Math. Soc. 50 (1985), 1–26.

[D4] S. K. Donaldson, *Infinite determinants, stable bundles and curvature*, Duke Math. J. 54 (1987), 231–247.

[G-T] D. Gilbarg and N. S. Trudinger, "Elliptic partial differential equations of second order", second edition, Springer-Verlag, Berlin and New York, 1983.

[H] N. J. Hitchin, *The self-duality equations on a Riemannian surface*, Proc. London Math. Soc. (3) 55 (1987), 59–126.

[K] S. Kobayashi, "Differential geometry of holomorphic vector bundles", Publ. Math. Soc. Japan, Iwanami Shoten and Princeton Univ., 1987.

[N-S] M. S. Narasimhan and C. S. Seshadri, *Stable and unitary vector bundles on a compact Riemann surface*, Ann. Math. 82 (1965), 540–564.

[Sim] C. Simpson, *Constructing variation of Hodge structure using Yang-Mills theorey and applications to uniformization*, J. Amer. Math. Soc. 1 (1988), 867–918.

[Siu] Y.-T. Siu, "Lectures on Hermitian-Einstein metrics for stable bundles and Kähler-Einstein metrics", Birkhäuser Verlag, Basel-Boston, 1987.

[U-Y1] K. Uhlenbeck and S.-T. Yau, *On the existence of Hermitian-Yang-Mills connections in stable vector bundles*, Comm. Pure Appl. Math. 39 (1986), S257–S293.

[U-Y2] K. Uhlenbeck and S.-T. Yau, *A note on our previous paper: On the existence of Hermitian-Yang-Mills connections in stable vector bundles*, Comm. Pure Appl. Math. 42 (1989), 703–707.

3

Stability of Vector Bundles on Surfaces and Curves

FEDOR A.BOGOMOLOV

Steklov Institute of Mathematics, Moscow, Russia

The main purpose of this article is to compare the properties of stable bundles on surfaces and of their restrictions on the curves. We consider the case of smooth projective surface V and stable vector bundle E, where stability means that E is stable in respect with some polarisations h on V.In fact we shall assume E to be stable in respect with some open subdomain U_E in the natural real envelope $Pic_R V$ of the group $PicV$. The main theorem of the article shows the existence of a subdomain U_E^s in U_E with the property that for any curve X the restriction of the bundle E on X is stable if the class of X lies in the the subdomain U_E^s . The curve X has not even to be smooth, but instead we need the quotient of the restriction of E on X by the destabilising subsheaf to be locally free.

The subdomain U_E^s can be precisely described inside U_E by two parameters: the dimension of E,denoted as r and the discriminant of E defined by the formula $\delta(E) = ((r-1)/2r)c_1^2(E) - c_2(E)$.

A connected component of the intersection of U_E with any two dimensional real subspace in $PicV \otimes R$ has a structure of cone,whereas for U_E^s it is an intersection of two domains. One of them has a hyperbolic shape with the asymptotic boundary lines on the boundary of U_E.The boundary curve is described then by equation of degree four in linear coordinates on the corresponding real plane with parameters dependent on the plane. Another one is an intersection of halfspaces ,with boundary hyperplanes defined by nonzero elements with trivial square in$PicV$.For precise description see theorems 2.3 and 3.4.

If U_E has relatively compact base inside the quotient of the cone $X^2 > 0$ by scalars then the difference between U_E and U_E^s contains only finite number of the elements in $PicV$.

Most of the ideas and computations of this article were in fact already contained in the article [B1].The main new result concerns the case of general bundle of dimension more than two.All the rest is a better arranged or slightly generalised version of the results from [B1] .

The problem above was also considered by several authors but most of the results obtained were concerned with stability of the vector bundles being restricted on generic curve or subvariety.

The method proposed in [B1] and considered in this article proves stability of the restriction for any curve with the class in U_E^s. In fact it gives the effective estimates for the level of stability of the restricted bundle. The main reason to write this article was to recall the method created in [B1],since I realised after several conversations that it is unknown to the specialists.

I was also inspired by the talk of P.Kronheimer at the Tokyo Conference on Algebraic geometry,where he used the transformation of antiselfdual connection along the riemannian surface in four-dimensional smooth manifold.

Though he was working with metrics and ASD connections the general flavour of his approach sounded strikingly similar to the method considered in [B1]. I hope this article will stimulate the attempts to understand the connection between his approach and purely algebrogeometric computations presented here.

The article contains four paragraphs.In the first one I considered the notion of stability based on the results of invariant theory for unstable points considered in [B1].

The second paragraph contains the results concerning two-dimensional case and mostly is a rearranged version of the results from [B1].

The third paragraph contains the stability criterium for the restrictions of bundles of any dimension.This is the main new result of the article.

I put also few concluding remarks into §4 concerning the immediate applications of the results obtained. I mentioned only few of them and did not give many details having a hope to write a detailed version later on. The list of references is as minimal as possible.I don't mention some related results in order to concentrate on the main subject.

The article reflects the talk given at Taniguchi conference in december 1990 and I want to thank the organisers of this conference (especially professors Mabuchi and Murakami) for the invitation and very stimulating atmosphere during the conference. This text was completed during my stay at Warwick University and I want to thank the members of the department of mathematics and especially Miles Reid for hospitality.

§1

In this paragraph I shall consider the notion of stability for the vector bundles on projective surfaces. I recall first that for a vector bundle on a smooth curve stability of the bundle E means the following inequality for any proper subbundle F of E: $\det F/dimF < detE/dimE$. In higher dimensions we have two differences:
1) the rank of the Picard group of the variety can be greater than one,
2) we must consider simultaneously torsion free subsheaves of the sheaf $O(E)$ and they don't coincide with the sheaves of sections of vector bundles on a variety.
Let us define first the group $Pic_R V$ for the projective surface V (or variety of any dimension as tensor product $PicV \otimes R$. This is a real vector space containig the quotient group $PicV/Pic^0 V$ as a sublattice.It is a subspace of the group $H^2(V, R)$ and the intersection of cocycles induces on $Pic_R V$ a hyperbolic quadratic form. We denote as K_P the convex envelope of the elements in $Pic_R V$ corres ponding to the polarisations of V in $PicV$.The cone K_P will be called further the polarisation cone.

The set of points in $Pic_R V$ with $x^2 > 0$ is divided into two cones and we shall denote as K^+ the one containing K_P as a subcone. We shall denote respectively as K_e the subcone of $Pic_R V$ generated by the elements corresponding to effective divisors on V. We shall consider only convex cones and the cone generated by a_i coincides with the elements of type $\Sigma\lambda_i(a_i)$ where all λ_i are nonnegative numbers and not all of them are zero.

As usual putting bar on the top of the notation will mean that we consider the closure

of the cone.The minus in front of the notation will mean we take the cone consisting of the elements $-x$ for all x in the cone. There is a natural sequence of the cone's imbeddings:
$K_P \subset K^+ \subset K_e$

and by Nakai- Moisheson theorem this sequence of imbedded cones is self dual on the open parts of the above cones. It is due to the fact that an element h of $PicV$ with the product
$(h, x) \geq \epsilon \deg(x)$ for any element x in K_e corresponds to the polarisation of the surface V.

Definition. (Mumford-Takemoto) We shall call the vector bundle E on a projective surface V to be stable in respect with an element h in K_P if $(c_1(F)/dimF)h < (c_1(E)/dimE)h$ for any proper torsion free subsheaf F of the locally free sheaf $O(E)$.

Here $dimF$ means the dimension of the fiber F/m_xF for the generic point x of a surface V.

Now I would like to define two special subcones in Pic_RV corresponding to the bundle E.

Definition $C_E(1)$. Consider for any locally free subsheaf F in $O(E)$ with $\dim E/F \geq 0$ the element $(c_1(F)/dimF) - (c_1(E)/detE)$ and define $C_E(1)$ to be the cone generated by these elements in Pic_RV.

It is clear that $C_E(1)$ does not change if we tensorise E with any linear bundle. In order to define another cone associated with E I recall that there exists a principal homogeneous bundle G_γ over V with G being an algebraic reductive group and such that E is associated to G_γ under some representation of the group G. Moreover we can choose G canonically.

To do this I recall the notion of semistable tensor for the reductive group G. If C_ρ is a linear representation of G then the element s of this representation is called semistable if the closure of G-orbit of this element contains the closed orbit which has dimension more than zero and less than the dimension of the generic orbit.

For any linear representation ρ of the group G we can define the bundle E_ρ associated to this representation. If there is a section semistable at some point,then this section is semistable everywhere and the orbits of it's values at different points are attached to the same closed orbit of G in C_ρ. Then by applying D.Luna slice-etale theorem (as in [B1]) we can reduce the structure group of the bundle E to a reductive subgroup of strictly smaller dimension. Thus there exists the reductive structure group G_E of the bundle which does not have semistable tensor sections in respect with the semisimple part of the connected component of G_E. It is clear that such group has uniquely defined connected component,and in fact it is uniquely defined for the fixed bundle E. Indeed the quotient group is described by the action on the tensor invariants for the connected subgroup and thus by locally constant but not constant stable tensor sections for E.

The character of the group $GL(dimE)$ defining the linear bundle $detE$ defines also the character of the subgroup G_E denoted in the same way. Thus for any irreducible representation ρ of the group G_E we can define a rational number $w(\rho)$ corresponding to the weight of the character $detE$ in ρ.

Definition $C_E(2)$. For any tensor bundle E^ρ and any one-dimensional subsheaf L of E^ρ we consider an element $c_1(L) - w(\rho)det(\rho)$ of the space $Pic_R V$. We denote by $C_E(2)$ the cone in $Pic_R V$ generated by the elements of this type.

Any torsion free subsheaf F of E with nonzero-dimensional quotient defines a subsheaf of dimension one in the exterior power of E of degree equal to $dim F$.This subsheaf contains the image of $det F$ as a subsheaf. Therefore for any element f of $C_E(1)$ we can find s in $C_E(2)$ with $s - f$ in $-K_e$. It is easy to see that for any element k in the interior of K_e both cones contains $s - k$ together with s.

Thus the closure $C_E(2)$ in $Pic_R V$ contains $C_E(1)$. The following statement gives a more precise description of these cones.

Theorem 1.1. *The cones* $C_E(1), C_E(2)$ *satisfy the following properties:*
both of them contain the interior part of the cone $-K_e$
they coincide outside of the closure $-K_e$
they are closed and locally simplicial outside of the closure of $-K_e$.

Proof. The first property follows from the fact that E contains some subsheaves of smaller dimension as the argument above shows. In order to proof the theorem we have to show that any relatively compact subdomain in the complementary of $-K_e$ in $Pic_R V$ contains only finite number of the boundary points of both cones $C_E(1), C_E(2)$ and the corresponding sets coincide . For any open relatively compact subset U in $Pic_R V \backslash (-K_e)$ we can find a polarisation h with $1 \leq h(u) \leq 2$ for any $u \in U$. It follows from the duality between the cones K_P and K_e described above. For any ample curve X on V we have only finite number of possible values for the degree of the positive subbbundles of the restriction of the bundle E on X. Thus any conic subdomain in $Pic_R V$ with relatively compact projection into the $(Pic_R V \backslash (-K_e))/R*$ will contain only finite number of the elements of $C_E(1)$ and therefore the cone $C_E(1)$ is closed.
I recall that by the theory of unstable tensors any unstable linear subbundle in E_ρ can be decomposed into a sum of the elements from $C_E(1)$ and $-K_e$. On the other hand any stable linear subbundle can be expressed as a sum of $det E_\rho$ and the element of $-K_e$. The determinants of the irreducible representations corresponds to the characters of the group G_E .
In the same way any linear sheaf L of unstable tensors can be imbedded into some special linear sheaf on V connected with the characters of parabolic subgroups of G_E. Namely any tensor bundle E_{chi} is constructed by some positive character χ of a parabolic subgroup P_χ of G_E. The vectors of dominant weight in the corresponding representation constitute a subfibration over the surface V.The fiber of it is a cone of dominant vectors.

The projectivisation of this conic fibration is an associated fibration with G_E/P_χ as a fiber. There is a linear bundle $L_{-\chi}$ on it corresponding to the dual character and the conic fibration is obtained by fiberwise contraction of the zero section of $L_{-\chi}$ to the surface V.

By the theory of G-equivariant map for unstable orbits we can map L into some conic fibration described above . It gives the equality
$L = s^*(L_{-\chi}) - k$.

Here s is rational section of the fibration above with the G_E/P_χ as a fiber and k is an element of K_e corresponding to the zero divisor of the bundle map. The section s is regular out of finite set of points .

The generators of $C_E(1)$ correspond in the same way to the positive characters of the parabolic subgroups of $SL(dimE)$,since parabolic subgroups in the latter case correspond to the flags of subspaces.

Therefore the theorem above will follow from the following lemma about the imbedding of reductive groups.

Lemma 1.2. *Let H, G be two reductive affine groups , $H \subset G$. Then any parabolic subgroup of H is contained in a parabolic subgroup of G, and some integer multiple of any nonnegative character for the Borel subgroup B_H of H is induced from some nonnegative character of Borel subgroup of G containing B_H.*

Remark. I think it is well known fact,but it is easier to give a proof than to find a reference.

Proof. Let us take any parabolic subgroup P of H and consider some positive character χ of P. The character defines an irreducible representation of the group H with the set of dominanant weight lines parametrised by H/P. Consider the imbedding of the maximal torus T_H of H into the one of G denoted as T_G. Now we can extend some multiple of the character χ on the whole T_G via orthogonal decomposition of the corresponding spaces of toric characters. I recall that for the case of simple Lie algebras we have a unique metric on the set of characters invariant under the Weil group $W(G)$ of the connected component of G.

The newly obtained character χ' of T_G has the same length as it's restriction on T_H.

Let us take now an irreducible representation R of G corresponding to the extended character. The representation of H considered above is a linear subspace in the new one.

The nontrivial characters in the decomposition of R over T_G are all contained in the convex hull of the orbit $W(G)\chi'$. Therefore the dominant vector in the representation of H are also dominant vectors of the corresponding representation of G. Thus there exists a parabolic subgroup P' with $H/P \subset G/P'$,but then P is a subgroup of P'.

Since the basis of positive characters is induced from subgroups P with $PicH/P = Z$ the lemma is proved.

If the quotient of G_E by it's connected component is nontrivial ,then the structure group of the bundle becomes connected on the finite unramified covering of V and we prove the statement of the theorem there first. The difference between characters of the group and it's connected component implies the existence of a direct deccomposition of the bundle on the covering above.

It is easy to see then that the linear bundles in $C_E(2)$ emerging on the covering differ from the elements in $C_E(2)$ by torsion elements in $PicV$ and therefore create no difference in $Pic_R V$.

The same is true for $C_E(1)$ This finishes the proof of the theorem

I want to remark also that in the case of absolutly unstable bundles ,or equivalently when $C_E(1) = C_E(2)$ we can define a canonical parabolic structure group P for the bundle E. Unfortunately it is well defined only outside of codimension two in general.

To do this let us take all unstable tensor sections.If we identify fibers over some point $x \in V$ with tensor representations of G_E then the values of this tensor sections constitute a linear tensor subalgebra of the whole tensor algebra.

There exists then a minimal parabolic subgroup P_x in G_E with the property that the closure of the orbit Ps_x contains 0 for any unstable tensor section. This groups vary in a natural way with x and we can consider the family P_x as a minimal destabilising subgroup of the group G_E defined over the field of functions on V.

§2

As a corollary of the theorem 1.1 we can deduce that if $C_E(1)$ does not intersect the closure of K_e then there exists a domain of stability of the bundle E.This is a subdomain in the polarisation cone K_P denoted further as U_E. The bundle E is stable in respect with any element of the domain U_E.

It is a conic subdomain that means it is invariant under homothetics in $Pic_R V$. I want to show now the way to describe a subdomain U_E^s in U_E with the property the restriction of E is stable on any curve with the class in U_E^s. In this paragraph we shall consider the case of two-dimensional bundles. The following lemma from [B1] will be essential because it measures the instability level of a bundle with positive discriminant.

Lemma 2.1. *Suppose that E is a two-dimensional bundle on a projective surface with positive discriminant*
$\delta(E) = (c_1^2/4) - c_2)$,*then E contains a torsion free subsheaf L with E/L of dimension one outside of finite number of points and*
$c_1(L) - 1/2(c_1(E))$ *in* K^+ ,
$(c_1(L) - 1/2(c_1(E)))^2 \geq \delta(E)$.

We know that E is unstable and therefore contains a subsheaf L which destabilizes the bundle E. In particular this means $c_1(L) - 1/2(c_1(E))$ does not lie in $-K^+$. Since L is a subbundle of E outside of finite set of points the following inequality holds for the second Chern class of E:
$c_2(E) \geq (c_1(L))(c_1(E) - c_1(L))$.

It becomes an equality if L is a subbundle of E everywhere on V . In general the difference is a sum of some positive integer numbers distributed on the singular points of the subsheaf L. Each of them depends only on the local behavior of the subsheaf L.

If we invert the signs on the both sides of inequality above and then add $1/4(c_1(E))^2$ to both of them we shall obtain the inequality of the lemma for L :
$(c_1(L) - 1/2(c_1(E)))^2 \geq 1/4c_1(E)^2 - c_2(E)$. Since the element above does not lie in $(-K^+)$ and has positive square it lies in K^+.

Remark. We have proved in fact that the statement of the above lemma holds for any onedimensional subsheaf in E if it is a subbundle outside of finite number of points and its first Chern class does not contain in $-K^+$.

Remark. I will assume in all further computations of this paragraph $detE$ to be trivial.It makes computations easier but does not change the result.To extend them for any bundle we have to apply them to the virtual bundle $E \otimes -detE$.

Let now X be a curve with the class in U_E and the restriction of E on X to be unstable .Thus E_X contains one-dimensional subbundle F_X of positive degree. For all further considerations we shall assume the quotient sheaf $O(E/F)_X$ to be locally free on X.

The quotient map of sheaves $O(E) \to O(E/F)_X$ is defined. The kernel of this map is a locally free sheaf which will be denoted further as $O(E')$ and the corresponding vector bundle on the surface V will be denoted as E'. I would like to compute now the Chern classes of E'.

Lemma 2.2. *The Chern classes of E' are described by the formulae:* $c_1(E') = -X$ *and* $c_2(E') = c_2(E) - FX$ *where FX means the degree of one-dimensional subsheaf F on the curve X.*

Proof. Since the map of E' into E degenerates exactly on the curve X and the rank of the image on X is one we obtain $c_1(E') = -X$ There are different ways to compute the second Chern class ofE' but I shall use here the one based on the additivity of the Euler characteristics.

Namely $\chi(E) = \chi(E') + \chi(E/F)_X$. By the Riemann-Roch formula $\chi(E/F)_X$ is equal to
$1 - p_a(X) + deg(E/F)_X$ The arithmetical genus $p_a(X)$ is given by the formula :
$1 - p_a(X) = (-1/2)X(K + X)$ and $deg(E/F)_X = -FX$.
It gives $\chi(E/F)_X = (-1/2)X(K + X) - FX$, where K is canonical class of the surface V. For two dimensional vector bundles we have Noether formula :
$\chi(E') = -c_2(E') + 1/2X^2 + (1/2)KX + 2\chi(O)$ and
$\chi(E) = -c_2(E) + 2\chi(O)$ since $c_1(E) = 0$. Applying the equality above and cancelling the terms with opposite signs we obtain:
$-c_2(E) = -c_2(E') - FX$ and the formula for $c_2(E')$ described in the lemma follows..

Corollary. The discriminant of the bundle E' is equal to $(1/4)X^2 - c_2(E) + FX$ and thus positivity of the above expression implies instability of the bundle E'.

Assume now that the discriminant of E' is positive and FX is non negative We shall obtain then by lemma 2.1 that E' contains one dimensional subsheaf L with $(c_1(L) + 1/2X)^2 \geq \delta(E')$ and $c_1(L) + 1/2X$ to be an element of K^+. In other words we have inequality:
$L(L + X) + (1/4)X^2 \geq (1/4)X^2 - c_2(E)$ where I denoted by L the class $c_1(L)$.

Cancelling $(1/4)X^2$ on the both sides of the equality we obtain
$L(L + X) \geq -c_2(E)$.

Let us analyse this inequality.If X lies in the positive cone K^+ then $L + X$ is also a positive element in the the group $1/2PicV$ and we can estimate it from below by another positive element $L + (1/2)X$. The map of L into E may have zero rank at the generic point of the curve X but then E would contain the subsheaf $L + X$ thus being unstable on V in respect with any polarisation. Therefore the rank of the map of L into E is one at the generic point of X and we can wright the following inequality using the arguments of lemma 2.1: $c_2(E) \geq -L^2$.

Assume now that X is inside the stability domain U_E for the bundle E and define the characteristic of the depth of the element X inside U_E.

For any Y ,X in K^+ the notion of angle can be naturally defined by the formula $1 - tan^2\varphi(X,Y) = (X^2)(Y^2)/(XY)^2$.

The expression above describes also invariantly defined hyperbolic tangents $th(\varphi)$.

For any $X \in U_E$ and W orthogonal to X we can define a halfplane in $Pic_R V$, containing both of them and $-X$.This halfplane intersects the domain U_E by the cone with one side corresponding to X.Consider the function $t_X W$ to be the tangents of the angle between X and another side of the cone above. The value of $t_X W$ is equal to the maximum of $tan\varphi(X,Y)$ for Y in the intersection of the halfplane defined by X, W with U_E. It is a function on a sphere of the vectors W with $W^2 = -1, WX = 0$.

Definition. Define for a curve X in the stability domain U_E the number t_x to be the minimum of the function $t_X W$ if $dim Pic_R V > 1$.

In the case when $Pic_R V$ is one-dimensional we define t_X to be 1 .

The number introduced above measures the distance of the direction defined by the element X in K^+ from the boundary of the domain U_E.It is more then zero,since X is inside U_E and the sphere of vectors W is compact. On the other hand $t_X \leq 1$ and $t_X = 1$ only if U_E coincides with the cone K^+, as it happens in the case $Pic_R V = R$

Theorem 2.3. *Two-dimensional bundle E with trivial first Chern class has stable reduction on any curve X inside the stability domain U_E with*
$X^2 > 4c_2^2/(t_X)^2$ and
$RX \leq c_2(E)$ for any nonzero element R of $(1/2)PicV$ with $R^2 = 0$ lying in the closure of K^+.

Proof. Let us consider the above formulae:
$L(L+X) \geq -c_2$ and $L^2 \geq -c_2$. Here and in the further computations I denote by c_2 the class $c_2(E)$ and consider it as a positive number. If $L^2 > 0$ and thus L is in $-K^+$ then $L(L+X)$ is negative and
$-(L^2 \times (L+X)^2 \geq -(L(L+X))^2$ .On the other hand $L^2 \geq 1/4$, since L is an element of the sublattice $1/2PicV$ in $Pic_R V$. We know also that
$(L+X)^2 \geq 1/4X^2$.

Thus $L(L+X)$ in this case is negative number estimated from above by $(-1/2)|X|$.
Here $|X|$ is a positive integer with $|X|^2 = X^2$.
That means $(1/4)X^2 \leq c_2^2$ .It proves the theorem in the case of $Pic_R V = R$.

Consider now real subspace M in $Pic_R V$ generated by L and X and the sublattice generated by this two vectors.. The determinant of this sublattice is negative number lying in the set $(1/4)Z \subset R$ and equal to $(X^2)(L^2) - (LX)^2$.

If we take the basis X, W in M with $W^2 = -X^2$ and $XW = 0$ then we can describe the intersection of the domain U_E with M by the formula: $X' \in M \cap U_E$ if X' is equal to $aX + bW$ with $-t_X < b/a < t_X$.

Therefore in the expression for L in the same basis we obtain $L = -dX + cW$,where d is a positive number and
$0 \leq d \leq c_2/(t_X X^2)$. It follows from the inequality
$-c_2 \leq LX < 0$ (since L^2 considered to be negative). The negativity of the product $L \times (aX + bW)$ under the above conditions on a,b gives $cb < ad$ or for $b = at_x$ we have $ct_X < d$.

L^2 can be expressed now by the formula :
$L^2 = (LX)^2/X^2 - (LW)^2/X^2$,or in the coeficients as $(d^2 - c^2)X^2$. It can be estimated from below by the second term,but
$-c_2/t_x \leq LW \leq c_2/t_X$ and thus
$L^2 \geq -c_2^2/t_X^2 \times X^2$. By assumption $L^2 \leq -1/4$ and therefore
$X^2 \leq 4c_2^2/t_X$.

Consider the remaining case $L^2 = 0$.
We have $LX \geq -c_2(E)$ and since L is nontrivial by stability assumption on E we obtain

the second type of inequalities defining subdomain U_E^s in U_E. This finishes the proof of the theorem

Remark. As it was pointed out to me by Maruyama the semistable bundles can actually have unstable reduction on the curves of appropriately big degree. The difference comes from the fact that the subbundle L above has to have negative degree on X. If we let LX to be zero then no contradiction will occur from the above considerations.

I want to point out that we used in the proof the integrity of quadratic intersection form on the group $PicV$.In case of dimension two bundle we actually had to consider the halfinteger elements and therefore obtained the coeficient 4 in our estimates.

Therefore an additional information about the intersection form on the Picard group of the given surface can improve the estimates. In the computations of the theorem 2.3 we also have used only halfinteger elements only to tighten the domain U_E^s. We can increase it using instead of constant t_x some more complicated function on the normal sphere to the element X. I have found it difficult to improve the above results in general,but under some additional information on the bundle and the surface it makes possible to obtain better estimates for U_E^s.

§3

In this paragraph I will extend the results of §2 to the bundles of any dimension. In order to apply the same kind of arguments as before we shall need an analogue of the unequality between Chern classes of a bundle and a subbundle. Let E be a vector bundle of dimension r and F be a subsheaf of E which is torsion free and a proper subbundle of E outside of finite number of points on a smooth surface V. In order to simplify the formulae we shall assume further that $c_1(E) = 0$. The formulae in general case can be obtained by twist with a virtual linear bundle $(1/dimE)detE$.

In this case the first Chern classes for F and E/F are well defined and we have equality:
$c_1(F) + c_1(E/F) = c_1(E)$. There exists a surface V' ,obtained from V by a sequence of blowing up points and such that F becomes locally free on V'(see for the most general formulation the theorem in [M1]). Moreover if we denote by p regular projection of V' onto V then the image ofp^*F in p^*E can be imbedded into a subbundle F' of p^*E of the same dimension. The quotient F'/p^*F is a sheaf concentrated on the preimage of the finite number of points of V. On V' we have an equality:
$c_2(E) = c_2(F') + c_2(E/F') - c_1^2(F')$. The class $c_1(F') = c_1(F) + \Gamma$, where Γ has a support on the preimage of finite number of points in V. Therefore
$c_1^2(F') = c_1^2(F) + (\Gamma)^2$,but Γ has nonpositive selfintersection. Thus if we define $c_2(F)$ as $c_2(F')$ the following inequality holds :
$c_2(E) \geq c_2(F) + c_2(E/F) - c_1^2(F)$.

Warning. The definition of the second Chern class of F given above depends on the imbedding of the sheaf F into locally free sheaf. It is easy to see that if the imbedding is given, then it does not depend on the surface V' where F' is defined.

We shall work mostly with discriminant of the bundle E which in our case coincides with $-c_2(E)$. I recall that in general the discriminant of the vector bundle of rank r is given by the formula :
$\delta(E) = ((r-1)/2r)c_1(E)^2 - c_2(E)$. Extending this definition of the discriminant to the

torsion free subsheafs of locally free sheaves we confront with a problem that it depends on the imbedding of the sheaf.

However the following proposition shows that the discriminant only increases with the increasing singularity of the map of the sheaf into locally free sheaf.

Lemma 3.1. *Let F be some locally free sheaf on a smooth surface V and suppose that F is also realised as a subsheaf of a locallyfree sheaf E with the quotient being locally free outside some finite set of points.Let F' be a locally free subsheaf of p^*E defined above with F'/F being torsion .*
Then $\delta(F') \leq \delta(F)$.

I want to remark first that since we consider the case of smooth surfaces any torsion free sheaf L has a canonical imbedding into some locally free sheaf with the quotient sheaf concentrated on a finite set of points. This corresponds to standard imbedding of the torsionfree sheaf into its reflexive envelope. The notion of second Chern class can be also defined for any sheaf via Riemann-Roch formula through the Euler characteristic of the sheaf. In this case second Chern class of the reflexive sheaf is smaller than the one for the locally free sheaf,what corresponds to the increase of the discriminant, while passing from the singular sheaf to its reflexive envelope.

Proof. Consider now the case of the map of locally free sheaf F into E. We have to understand only local change of the numbers
$c_1^2(F), c_2(F)$ in the neighborhood of the exeptional divisor,which contains smooth rational curves. Thus we can assume F to be isomorphic to a sum of the linear bundles $O(\Gamma_i)$,where each Γ_i is an effective combination of the rational curves. In this set up, the local formula for $c_2(F')$ will be
$\Sigma\Gamma_i\Gamma_j$ for $i > j$ and $c_1(F') = \Sigma\Gamma_i$.

Thus the difference between discriminants will be equal to
$(1/2)(\Sigma(\Gamma_i)^2) - (1/2r)(\Sigma\Gamma_i)^2$. Since the intersection form on the system of rational curves above is negative definite the preceding expression has nonpositive value for any combination of curves Γ_i. That means the discriminant of F' is bounded from above by the discriminant of F.

I want to prove first a weak analogue of lemma 2.1 for the bundles of big dimensions (already contained in [B1]).

Lemma 3.2. *If $\delta(E) > 0$ then E contains a locally free subsheaf L which is a proper subbundle of rank l outside of finite number of points on the surface and $c_1(L) - (l/r)c_1(E)$ to be the element of K^+.*

We know that E is unstable for any polarisation on the surface V. We can also assume that the lemma is true for any bundle of smaller dimension.

Suppose that the proposition is not true for E and obtain the contradiction. For any subsheaf F of E we can define a subdomain of the cone K^+ where this subsheaf is destabilising and the union of such domains denoted as D_F will cover the whole cone K^+. Suppose that the domain D_F is maximal in the sense that there are no domains $D_{F'}$ containing D_F and D_F does not coincide with the whole cone K^+. Let us prove that it cannot occur. Consider the subsheaf F in E,corresponding subbundle F' in p^*E and the bundle p^*E/F'.

By the formula for the Chern classes we have that
$\delta(E) = \delta(F') + \delta(p^*E/F') + ((1/2f) + (1/2(r-f))) \times c_1^2(F)$
Here f means the dimension of F.

Since we know that the first Chern class of F does not lie in $-K^+$ there are two possibilities: either $c_1^2(F) > 0$ and F is a subsheaf we wanted to construct,or $c_1^2(F) \leq 0$ and at least one of the bundles $F', p^*E/F'$ has discriminant greater than zero. We have to consider only the second case. Suppose that $\delta(F) > 0$. Since the dimension of F is strictly less than dimension of E we can find a subsheaf L of dimension l in F' with $c_1(L) - (l/f)c_1(F)$ being in K^+. Therefore the domain D_L in K^+ is strictly greater than D_F and we obtain a contradiction with the assumption on D_F. If the discriminant is greater than zero for p^*E/F than it contains a subsheaf R with similar property and we obtain contradiction by taking the sheaf $F + R$.

Remark. If L is a subsheaf of p^*E and $c_1(L)$ is contained in K^+ then L also defines a subsheaf L' of E and the class $c_1(L')$ is also contained in K^+ for the surface V. Moreover $c_1^2(L') \geq c_1^2(L)$. It follows from the fact that corresponding Chern classes differ by the combination of the divisors with negative intersection matrix.

The following proposition is a quantative version of lemma 3.2. The estimate obtained in this lemma makes it possible to extend the theorem 2.3 to the bundles of any dimension.

Lemma 3.3. *Let E be a vector bundle of rank $r > 1$ on smooth projective surface V with $c_1(E) = 0$ and $\delta(E) = -c_2(E) > 0$ .Then E contains a subsheaf F,which is a subbundle outside of finite set of points, $c_1(F) \in K^+$ and*
$c_1^2(F) \geq (2f^2/r(r-1))\delta(E)$.
Here f means the dimension of F.

Proof. We shall prove it by induction on the dimension of E.Thus we assume that the lemma is true for all dimensions smaller than r. In fact it is evidently true for dimension one and was proved for dimension two in the lemma 2.1

Let F be a torsion free subsheaf in E which is a subbundle outside of finite set of points and $c_1(F) \in K^+$.

Suppose also that the ratio $c_1(F)^2/f^2$ is maximal among all such subsheaves of E with $c_1 \in K^+$ but $c_1^2(F)/f^2 < 2/r(r-1)\delta(E)$.

I want to construct another sheaf with greater ratio above and thus prove that the assumptions on F are contradictive. This will prove the lemma.

We can assume as before that F is a subbundle by changing V to V'. Consider the equality from lemma 3.2:
$\delta(E) = \delta(F) + \delta(E/F) + ((1/2f) + (1/2(r-f))) \times c_1^2(F)$.

The sheaf F has nonpositive discriminant,because otherwise we can find a subsheaf L in F with greater ratio $c_1(L)^2/(dim L)^2$ than the one for F.

Indeed if $\delta(F) > 0$ then there exists a proper subsheaf L of dimension l with $(c_1(L) - (l/f)c_1(F)) \in K^+$. That means in particular $(c_1(L)^2 > (l/f)^2 c_1(F)^2$. This contradicts the assumption on F. Thus $\delta(F) \leq 0$ and
$\delta(E/F) + (r/2f(r-f))c_1(F)^2 \geq \delta(E)$.

From the assumption on $c_1(F)^2$ we obtain that the second term on the lefthand side is smaller than $((r/2f(r-f))2f^2/r(r-1))\delta(E)$ or after reduction $(f/(r-f)(r-1))\delta(E)$.

Thus we have inequality
$\delta(E/F) > (1 - (f/(r-1)(r-f))\delta(E)$. The coefficient on the righthand side can be rewritten as
$((r-1)(r-f) - f)/(r-1)(r-f) = r(r-f-1)/(r-1)(r-f)$.

Thus we obtain the following inequality for $\delta(E/F)$
$\delta(E/F) > (r(r-f-1)/(r-1)(r-f)\delta(E)$.

That means E/F contains destabilising subsheaf L of dimension l and with
$(c_1(L) + (l/(r-f))c_1(F))^2 > (2l^2/(r-f)(r-f-1))\delta(E/F)$

Applying the above inequality for $\delta(E/F)$ we obtain the following coefficient at $\delta(E)$ on the righthand side :
$(2l^2/(r-f)(r-f-1))r(r-f-1)/(r-1)(r-f)$ and after the reduction
$2rl^2/(r-1)(r-f)^2$.

Thus we have the following inequality
$(c_1(L) + (l/(r-f)c_1(F))^2 > (2rl^2/(r-1)(r-f)^2)\delta(E)$

Let us consider the bundle $F + L$ and prove that
$c_1(F+L)^2/(f+l)^2 \geq (2/r(r-1))\delta(E)$ Indeed let us decompose this class into a sum of two elements
$c_1(L+F) = (c_1(L) + (l/(r-f)c_1(F)) + (r-l-f)/(r-f)c_1(F)$

Both of the elements above lie in K^+ and we have the following inequality inside this cone
$(X+Y)^2 \geq X^2 + Y^2 + 2|X||Y|$.

Thus if $c_1(F)^2 = 2a^2f^2/r(r-1)\delta(E)$ with $a<1$ we obtain the following coefficient at $\delta(E)$in the expression estimating $c_1(L+F)^2$ from below:
$(2/r(r-1)(r-f)^2)(l^2r^2 + 2afrl(r-f-l) + a^2(r-l-f)^2f^2)$

Consider the last part of the expression .It is a square of the number $lr + a(r-f-l)f$. This number is greater than $a(l+f)(r-f)$.Indeed the difference is
$lr + arf - af^2 - alf - alr - arf + alf + af^2$.

Cancelling the same terms with opposite signs we have
$lr(1-a)$ which is positive since $a < 1$.

But then the coefficient above is greater than
$(2/r(r-1)(r-f)^2)(a^2(l+f)^2(r-f)^2$ or after reduction $(2/r(r-1))a^2(l+f)^2$.

Thus $c_1(L+F)^2/(l+f)^2 > c_1(F)^2/f^2$. This proves that the assumptions on F were contradictive and finishes the proof of the lemma.

Corollary. Let E be a vector bundle on a surface with trivial determinant and $\delta(E) > 0$. Then E contains a subsheaf F (free outside finite number of points) with $c_1(F) \in K^+$ and $c_1(F) \geq 2(r-f)^2/r(r-1)\delta(E)$, where r is a rank off E and f is a rank of F respectively. It follows by applying the above lemma to the dual bundle E^* of E.

Remark. If dimension (rank) of the bundle E is equal to two,then the proposition above coincides with the estimate given by lemma 2.1.

Definition. We shall call a subsheaf L of dimension l in E with $c_1(L) - l/rc_1(E) \in K^+$ to be the subsheaf with maximal slope if $(c_1(L) - l/rc_1(E))^2/l^2$ has maximal value among all such relatively positive subsheaves in E

This notion makes it possible to describe an analogue of Harder-Narasimhan filtration for the bundles on surfaces.

We turn now to the proof of an analogue of the theorem 2.3 for the bundles of any dimension using the estimate above. Suppose as before,that we have a bundle $E', c_1(E') = 0, rankE' = r$,which is stable in a subdomain U'_E of the polarisation cone. Let X be a curve with its class containing in U'_E and assume that the restriction of E on X is unstable.Moreover we fix F'_X as a destabilising locally free subsheaf on X of the restriction E'_X. I would like now to describe precisely the subdomain of U'_E, where the class X cannot appear.

I shall use for X the same characteristics of it's position inside U'_E as were introduced in the second paragraph.

Theorem 3.4. *Let E' be a bundle described above and the restriction of E' on the curve X is unstable, while the class of X is contained in the stability domain of the bundle E'. Then X lies inside the union of domains defined by the following system of equations:*
$X^2 < A_f r^2 c_2^2(E')$ *and*
$RX \le B_f c_2(E')$ *for any nontrivial R of $1/rPicV$ with $R^2 = 0$ lying in the closure of* K^+.
The constants A_f and B_f depend only on r and dimension of the destabilising subsheaf F'_X.

Proof. Let F be the determinant of F' and E be the exterior power of E' of the degree $dimF'$. The bundle F is one dimensional and destabilises the restriction of E on X.Let N be the dimension of E . As before we can consider the locally free sheaf R which is a kernel of the projection map $O(E) \to (E/F)_X$ and compute its Chern classes and discriminant. They are : $c_1(R) = (1 - N)X$
$c_2(R) = c_2(E) - FX + ((N-1)(N-2)/2)X^2$,where FX denotes the degree of the bundle F on the curve X.
$\delta(R) = -c_2(E) + FX + ((N-1)/2N)X^2$.

Indeed the only part to compute is the coeficient at X^2, but it is equal to
$(N-1)^3/2N - ((N-1)(N-2)/2)$ or
$(N-1)(N^2 - 2N + 1 - N^2 + 2N)/N$,and after cancellation $(N-1)/2N$.

Thus for the destabilising subsheaf L of rank l with maximal slope we obtain the following inequality assuming $FX \ge 0$:

$(c_1(L) + X(l(N-1))/N)^2 \ge (2(N-l)^2/N(N-1))\delta(R)$, and $\delta(R) > 0$.

Notice that if the rank l is greater than 1 the image of L in E degenerates along X with multiplicity at least $l-1$. Therefore E contains a subsheaf L' of the determinant greater than $c_1(L) + (l-1)X$.
Now the above inequality in terms of the bundle L' is
$(c_1(L') + ((N-l)/N)X) \ge (c_1(L) + (l(N-1)/N)X)$ and therefore
$(c_1(L')(c_1(L') + (2(N-l)/N)X) + ((N-l)^2/N^2)X^2 \ge -(2(N-l)^2/N(N-1))c_2(E) + ((N-l)^2/N^2)X^2$.
The coefficients at X^2 on both sides coincide and therefore we obtain the inequality

$L'(L' + (2(N-l)/N)X) \ge -(2(N-l)^2/N(N-1))c_2(E)$.

I denoted by L' the element $c_1(L')$ in order to make the above inequality look more similar to the one considered in §2.
Now we shall easily obtain the same kind of inequalities to be satisfied by X.We have to use the same argument with the multiplication in the lattice $1/NPicV$ under the

assumption $L^2 \geq 0$. The intersection $L'X$ is negative because X lies in the stability domain U_E.Therefore in this case we obtain:
$L'^2 \geq -(2(N-l)^2/N(N-1))c_2(E)$ but we know that the difference $(L' + (2(N - l)/N)X - (L + l(N-1)/N)X$ is greater than $((l-1) + (2(N-l)/N - l(N-1)/N)X$ and this bundle is more positive than $((N-l)/N)X$. It can be estimated from below by $(1/N)X$ or in terms of the initial bundle E' by the maximum of the dimensions of exterior powers.

We can apply now the same kind of argument as in the case of dimension two bundle. The constants A_f, B_f can be precisely described by the following data:
the formula for $c_2(E)$ as a multiple of $c_2(E')$;
the lattice $(1/N)PicV$ and the quadratic intersection form on it;
the coefficient t_X introduced in §2 and characterising the position of X inside domain U_E.
In both cases the corresponding multiple of the coeficient depends on N and thus on the dimension of F.

Remark. It is easy to obtain the exact values for the coeficients in the theorem, but I don't do it here ,because I think that more elaborate version of this approach may significantly improve the resulting constants.

§4

I would like to make here several concluding remarks.The trick which was used to prove the above theorems can be applied in much more broad variety of problem

The technique developed above also shows the existence of the linear estimate of the degree for the subbundles of the stable bundle restricted on curve . Indeed, if in the above formula FX is a negative term but linearly bounded from below in respect with the degree of X then we can find analogous hyperbolic subdomain in U_E where X can not appear. Thus using the same formulae, but taking in account the degree FX we can estimate this number from above by a linear function of degree with negative coefficient in terms of the invariants $c_2(E), rankE, U_E$.
Another applications are related to the famous Donaldson's theorem,that any stable bundle E with $c_2 = c_1 = 0$ is obtained from the unitary representations.

Using the above formulae we can see that E has to have stable or semistable restriction on any curve with positive selfintersection but then there is a family of unitary representations of the fundamental groups of curves corresponding to the bundle E. Therefore we have a section in the family of the moduli varieties of stable bundles on curves in the pencil.The second Chern class of the vector bundle can be expressed then as a value of the multiple of relative anticanonical class over this section. The family of this moduli spaces is a nonlinear analogue of the family of intermediate jacobians for the Hodge structure variation. Therefore a kind of semipositivity statement for the relative anticanonical class together with natural flat connection on the family of moduli spaces would imply the Donaldson theorem.
The simplest case to consider is then a Kodaira surface with the pencil of smooth curves parametrised by compact base .
Triviality of the second Chern class would imply that the corresponding representation is constant and therefore our bundle is constructed by the representation of the fundumental group.

This scheme of proof has some technical problems . (In fact this approach to the description of the discriminant of vector bundle on surfaces was discussed already during ICM 1978, but since then nobody worked it out). A series of new results on the moduli spaces of vector bundles (Hitchin, Donaldson and others) makes more plausible the possibility to obtain the proof of the Donaldson's theorem along this line.

I also want to mention the application of the method considered above to the problem of rational curves on surfaces,but it demands another paper to be written.

In general applying the arguments above to the singular curves and configurations makes it possible to obtain various general position statements.

Bibliography

[B1] F.A.Bogomolov Holomorphic tensors and vector bundles on projective varieties Izvestya of Ac.of Sc.USSR ser.math v42 N6 1978 p.1227-1287

[M1] B.G.Moisheson Algebraic analogue of complex spaces Izvestya of Ac. of Sc.USSR ser.math.v.33 N1 p.174-238

4

Magnetic Monopoles and Topology

Peter J. Braam
Mathematical Institute
University of Oxford
Oxford, England

1 Introduction

This short paper is essentially a written version of the paper delivered at the Taniguchi symposium. We introduce magnetic monopoles in §2, continue with polynomial invariants in §3 and summarize some of their properties in §4. Details can be found in [B1], and for background we refer to [B2] and Donaldson-Kronheimer [DK]. NSF support is gratefully acknowledged.

2 Magnetic Monopoles

A magnetic monopole is a solution to a certain partial differential equation on a Riemannian 3-manifold. The key issue is that the solutions of this equation are parametrized by a nice space, usually a manifold with mild singularities, called the *moduli space of monopoles.* These moduli spaces carry a wealth of geometric and topological information concerning the underlying 3-manifold, which is largely unexplored. Here we will introduce topological invariants of the three manifold which are defined in terms of the topology of the moduli space of monopoles.

Let M be a Riemannian 3-manifold. We assume that M is complete and has ends of the form

$$\mathbf{R}_{>0} \times S_j$$

for a finite set of Riemann surfaces $S_j, j = 1, \ldots, n$. Furthermore, these ends are supposed to carry a complete Riemannian metric which is asymptotically approximately Euclidean [Fl1] or approximately hyperbolic [B2]. Let $P \to M$ be a, necessarily trivial, principal $SU(2)$-bundle. We define a configuration space

$$\mathcal{C} = \{(A, \Phi); A \text{ a connection on P}, \Phi \in \Gamma(M, P \times_G su(2)), ||F^A||_{L^2},\ ||d_A\Phi||_{L^2} < \infty\}.$$

Note that Φ is a skew adjoint endomorphism of

$$E \equiv P \times_G \mathbf{C}^2.$$

A *magnetic monopole* is an absolute minimum of the Yang-Mills-Higgs functional:

$$Y(A, \Phi) = \frac{1}{4\pi} \int_M -tr((F^A)^2) - tr((d_A\Phi)^2) dV,$$

and can alternatively be described as a solution to the Bogomol'nyi equation:

$$F^A = - * d_A \Phi.$$

In addition to the equations we impose certain boundary conditions, the so called *Prasad Sommerfeld boundary conditions*, which first arose in physics. The boundary conditions are asymptotic and describe the behavior at infinity on the ends $\mathbf{R}_{>0} \times S_j$. We require:

$$lim_{t\to\infty}|\Phi(t, s_j)| = m_j \in \mathbf{R}_{\geq 0},$$

that is, the length of Φ approaches a constant value on each of the ends. The constants m_j are called the *masses* of the monopole. If on an end $\mathbf{R}_{>0} \times S_j$ we have $m_j \neq 0$ then asymptotically the endomorphism Φ splits the bundle E into a sum of line bundles, the bundles of eigenvectors belonging to eigenvalue $\pm i m_j$ of Φ:

$$E_{|t\times S_j} = L_j \oplus L_j^*.$$

The second boundary condition states:

$$c_1(L_j^*) = k_j \in \mathbf{Z}$$

i.e. we restrict ourselves to configurations of given *charges*. When the charges and masses of a configuration (A, Φ) are fixed a partial integration in the action functional gives that:

$$Y(A, \Phi) = \sum 2m_j k_j.$$

The gauge group

$$\mathcal{G} \equiv C^\infty(M, SU(2))$$

acts on the configuration space $\mathcal{C}$ and the Yang-Mills-Higgs functional is invariant. The *moduli space* of monopoles on P is now defined as the space of minima of Y modulo the action of the gauge group. Given the boundary conditions it can be specified by

$$\mathcal{M}(m_j, k_j),$$

keeping in mind its dependence on the Riemannian metric on M.

We see immediately that a necessary condition for monopoles to exist is that

$$\sum m_j k_j \geq 0.$$

Some general existence theorems have been shown when all m_j are equal (Jaffe-Taubes [JT], Floer [Fl1], Braam [B2]) but it remains somewhat mysterious for which (m_j, k_j) monopoles exist, in particular when not all m_j are equal, see [B1]. A monople on $\mathbf{R}^3$ with charge $k = 1$ and mass m was discovered by 'tHooft and Polyakov and it looks like a particle of size $1/m$. The existence theorems use these known solutions on $\mathbf{R}^3$ to graft them into a backgound solution already present on M. In a rough first approximation a monopole $\mathcal{M}(m_j, k_j)$ could be thought of as k_j particles of size $1/m_j$ near the end S_j of M grafted into a flat connection, the vacuum.

When monopoles exist they come in moduli spaces. After a small perturbation of the Bogomol'nyi equations the moduli spaces are smooth manifolds provided that $dim(H^2(M; \mathbf{R})) > 0$ and $\sum m_j k_j \neq 0$. When one relaxes the condition on $H^2(M)$ singularities may appear at configurations with a non-generic stabilizer in $\mathcal{G}$; such configurations are referred to as *reducible monopoles*. This is very analogous to the situation one encounters with the anti-self duality equation in dimension 4. If all $m_j \neq 0$ the moduli space has the following dimension:

$$dim\mathcal{M}(m_j, k_j) = \sum 4k_j - (1 + b^2(M) - b^1(M)),$$

where b^i denote the Betti numbers of M.

A sequence of monopoles can fail to converge in $\mathcal{M}(m_j, k_j)$, but the failure of convergence is simple. It states that a number of 'monopole particles' can move off to infinity on the ends $\mathbf{R}_{>0} \times S_j$ and leave a monopole of lower charge (and the same masses) on M. One can actually assign disappearance points in the S_j to a suitable subsequence of monopoles. The precise theorem is that $\mathcal{M}(m_j, k_j)$ can be compactified by adding lower strata to $\mathcal{M}(m_j, k_j)$ which are subsets of

$$\mathcal{M}(m_j, k_j - l_j) \times \prod_j S^{l_j}(S_j).$$

Notice that *bubbling off*, the phenomenon that particles decrease their size to become a type of delta function, doesn't happen: the size of a monopole is not a parameter in the moduli space. This reflects the fact that the equations do not have conformal invariance. This is a significant difference with the geometry of instantons in dimension four where conformal invariance makes bubbling off an important aspect of the theory. It should be noted that considering the family of moduli spaces $\cup_{m_j} \mathcal{M}(m_j, k_j)$, on which the m_j are now functions, has proved to be very interesting. Compare Floer [Fl1].

We are now ready to discuss some topological applications of the moduli spaces.

3 Polynomial Invariants for 3-Manifolds

Analogously to Donaldson's definition of polynomial invariants for 4-manifolds one seeks to exploit the homology of $\mathcal{B} = \mathcal{C}/\mathcal{G}$. Inside $\mathcal{B}$ one finds the submanifold $\mathcal{M}(m_j, k_j)$, and this submanifold becomes canonically oriented upon orienting $H^1(M;\mathbf{R}) \oplus H^2(M;\mathbf{R})$. When $b^2(M) > 2$ one can show that considering $\mathcal{M}$ as a homology class and pairing it with certain cohomology classes on $\mathcal{B}$ is well defined and the outcome is independent of the choice of metric on M. This defines a topological invariant of M, the precise form of which turns out to be a polynomial on $H_2(M;\mathbf{Z}) \oplus H_1(M, \partial M;\mathbf{Z})$.

There is one point to be very careful with. The moduli space may contain reducible configurations. These cause complications in the cohomological machinery and it is carefully arranged for that this will not happen.

The cohomology classes in question are defined using a map μ:

$$\mu : H_2(M;\mathbf{Z}) \oplus H_1(M, \partial M;\mathbf{Z}) \rightarrow H^2(\mathcal{B}^*;\mathbf{Z}),$$

similar to the one introduced by Donaldson. Here $\mathcal{B}^*$ is the smooth part of $\mathcal{B}$ consisting of those pairs (A, Φ) for which the stabilizer in $\mathcal{G}$ is finite. The definition goes as follows. Any class in $H_2(M;\mathbf{Z})$ can be represented by a closed oriented surface $\Sigma \subset M$. We now couple the Dirac operator on Σ to the connections A on the $\mathbf{C}^2$-bundle E for $(A, \Phi) \in \mathcal{B}^*$. This gives a family of Dirac operators parametrized by $\mathcal{B}^*$ and a determinant line bundle $\mathcal{L}_\Sigma \rightarrow \mathcal{B}^*$. We define

$$\mu(\Sigma) = c_1(\mathcal{L}) \in H^2(\mathcal{B}^*;\mathbf{Z}).$$

To define $\mu(\lambda)$ for a class $\lambda \in H_1(M, \partial M;\mathbf{Z})$ we pick a curve $\lambda \subset M$ and consider the Hitchin operator:

$$\nabla_{A, \frac{\partial}{\partial s}} + i\Phi : \Gamma(\lambda, E_{|\lambda}) \rightarrow \Gamma(\lambda, E_{|\lambda}),$$

where s is a local parameter on the curve. Once more this defines a family of elliptic operators parametrized by $\mathcal{B}^*$. The first Chern class of the determinant index bundle $\mathcal{L}_\lambda \rightarrow \mathcal{B}$ defines $\mu(\lambda) \in H^2(\mathcal{B}^*;\mathbf{Z})$.

We now come to the definition of the invariants. Select charges and masses (k_j, m_j), and let $m = \frac{1}{2} dim \mathcal{M}(k_j, m_j)$. Choose m_1 surfaces Σ_n and m_2 curves λ_i for $m_1 + m_2 = m$ in suitable generic position with respect to each other and with respect to the 'boundary' surfaces S_j. Further, select generic sections s_i with $i = 1, \ldots, m$ of each of the determinant line bundles restricted to the moduli space:

$$\mathcal{L}_{\Sigma_n} \to \mathcal{M}(k_j, m_j)$$

and

$$\mathcal{L}_{\lambda_i} \to \mathcal{M}(k_j, m_j).$$

Denote by $z(s_j)$ the zero set of these sections. Let

$$\Phi_M(\Sigma_n, \lambda_i, s_1, \ldots, s_m, k_j, m_j) = \#(z(s_1) \cap \ldots \cap z(s_m)) \tag{1}$$

with # denoting an oriented count, using a choice of orientation of $H^1(M;\mathbf{R}) \oplus H^2(M;\mathbf{R})$.

Notice that we have tacitly assumed that the moduli space is contained in $\mathcal{B}^*$, i.e. that no reducible configurations are present. We shall now discuss this issue in somewhat more detail. A reducible solution to the monopole equations is a connection A on a line bundle $L \to M$, together with a function Φ on M such that

$$F^A = - * d\Phi.$$

The cohomology class $d\Phi \in H^1_c(M;\mathbf{R})$ is the image of the class $(m_j) \in H^0(\partial M;\mathbf{R})$ under the natural map $H^0(\partial M;\mathbf{R}) \to H^1_c(M;\mathbf{R})$. The Hodge star $* : H^1_c(M;\mathbf{R}) \to H^2(M;\mathbf{R})$ is an isomorphism, and the equation for a reducible monopole expresses that the lattice $2\pi i H^2(M;\mathbf{Z})$, which contains F^A, also contains the class $*(m_j)$. The following lemma should now be plausible:

Lemma 1 *If dim $H^2(M;\mathbf{R}) \geq 1$ then no reducible monopoles with non-zero curvature exist for a Baire set of metrics on M. If dim $H^2(M;\mathbf{R}) \geq 2$ then no reducible monopoles with non-zero curvature will occur in a generic one parameter family of metrics.*

There are other ways in which we can avoid reducible connections. The line bundle L satisfies $< c_1(L), S_j >= \pm k_j$ for all j. Notice that this is a strong condition on the set of charges. For example, when there are only two surfaces S_j we must have $k_1 \pm k_2 = 0$ for reducible connections to occur. This includes into our considerations the interesting case of complements of links of at least two components in 3-manifolds.

The situation with the flat reducible connections is less serious as the line bundles $\mathcal{L} \to \mathcal{M}(m_j, k_j)$ extend over these. Unless all masses are zero, the stabilizer of any reducible connection in the gauge group $\mathcal{G}$ equals $U(1)$.

Theorem 2 *Assume that*

$$dim\ H^2(M;\mathbf{R}) \geq 2$$

or that otherwise reducibles apart from the flat $U(1)$-connections do not appear in generic 1-parameter families of compactified monopole moduli spaces. Assume that not all m_j are equal, or that

$$\sum 2k_j \geq dim\ H^2(M;\mathbf{R}) + 2.$$

Choose a homology orientation.

Then $z(s_1) \cap \ldots \cap z(s_m)$ is finite. The function Φ_M is independent of the choices of the sections s_j, independent of the choice of generic metric g, and it depends only on the homology class of Σ_1 and is a linear function of this homology class. It is symmetric in the pairs (Σ_j, s_j). Consequently it defines a topological invariant of M.

This then allows us to define:

Definition 3 *Let k_j, m_j satisfy the condition of theorem 2, and suppose that no reducibles occur in generic 1-parameter families of compactified moduli spaces $\mathcal{M}(k_j, m_j)$*

because $b^2(M) \geq 2$ or otherwise. Select a homology orientation of M. The polynomial invariant Φ_{M,m_j,k_j} is the integer valued polynomial function on $H_2(M;\mathbf{Z}) \oplus H_1(M,\partial M;\mathbf{Z})$ derived from (1).

An immediate corollary is:

Proposition 4 *If M admits an orientation preserving diffeomorphism which reverses the homology orientation and preserves the numbers (k_j, m_j) attached to the S_j then $\Phi_{M,m_j,k_j} = 0$.*

Notice that we have defined infinitely many invariants for complements of links in 3-manifolds, when there are at least two components. Oriented links in a homology sphere define a homology orientation, by elementary algebraic topology.

The topology of the configuration space $\mathcal{B}^*$ is very rich and presents many more possibilities for detecting invariants. For example the following. One can extend our invariant to a 'super polynomial' with rational values on

$$H_1(M;\mathbf{Z}) \oplus H_2(M;\mathbf{Z}) \oplus H_1(M,\partial M;\mathbf{Z})$$

where the degree of elements in $H_1(M;\mathbf{Z})$ is 3. Elements of $H_1(M;\mathbf{Z})$ define submanifolds of the moduli space of codimension 3 by fixing the monodromy of connections along loops.

Another extension which does not seem to present much difficulty is to use the map:

$$R : \mathcal{A}(P)^{S^1}/\mathcal{G}(P)^{S^1} \to \mathcal{A}(P_{|\partial M})^{S^1}/\mathcal{G}(P_{|\partial M})^{S^1}.$$

New cohomology classes arise from pulling classes back from the space of connections on the boundary of M. These classes can be realized by fixing holonomy of connections on ∂M.

The most challenging possibility may be that of using $H_2(M,\partial M;\mathbf{Z})$. Through the μ map these lie in the cohomology of the fibre of R. They are difficult to realize by using Dirac index bundles.

It is also possible to use $SO(3)$-bundles. This opens up more possibilities as reducible connections can more easily be avoided. For example one may count flat $SO(3)$ connections on a bundle $P \to M$ when M is a complement of a two component link in S^3, and $w_2(P) \neq 0$. Likely the answer is the linking number, and the arguments in favour of this are as follows. When using a Heegaard splitting of M, as in the definition of the Casson invariant see [AM], one finds that this number equals the linking number of the components (Braam, unplished); this is well defined by the homology orientation of M. Similarly Floer [Fl2] finds this when counting connections directly on M. The statement 'likely' has to be inserted since it is not certain that the signs in these various ways of counting agree.

We are quite confident that a conjecture close to the following may be proved analogously to the proof of Donaldson's vanishing theorem for connected sums:

Conjecture 5 *Let M be a manifold satisfying the conditions mentioned in definition 3 for integers m_j, k_j. Let N be a compact, closed 3-manifold, and denote by $M\#N$ the connected sum. If N is a rational homology sphere then*

$$\Phi_M(k_j, m_j) = ord(H_1(N;\mathbf{Z}))\Phi_{M\#N}(k_j, m_j),$$

if N has $rk(H_1(N;\mathbf{R})) > 0$ then

$$\Phi_{M\#N}(k_j, m_j)_{|H_1(M,\partial M)\oplus H_2(M)} = 0.$$

When we take connected sums of manifolds with boundary this will only hold in much less generality. The relative one classes on connected sums could well behave in a complicated fashion under connected sums. For comparison see section 4.

4 Special cases of the polynomial invariants.

In this section we shall show the results of some computations of the coefficients of the polynomial invariant. The main results are that the polynomial is not zero, and, in fact, it grows rapidly with the charge. The manifold M will be $S^2 \times \mathbf{R}$; for slightly more general computations see Braam [Br1]. This manifold is useful in that it allows for a detailed study of the moduli space of monopoles through algebraic geometry. We shall omit the details here and just describe the results.

Notice that we have two ends in M so we have two charges k_0, k_∞ and two masses m_0, m_∞. As $H^2(M)$ is one dimensional we should keep one of the charges, k_∞ say, equal to 0 in order to have well defined invariants. When considering the situation with both charges non-zero the polynomial is well defined for any generic metric, but it may change when we vary the metric.

First let us look at the moduli space with $m_0 \neq 0$, $m_\infty = 0$. For an approximately hyperbolic metric on $S^2 \times \mathbf{R}$ it can be shown that the moduli spaces are empty unless $k_0 \geq m_0$; this is one of the mysterious existence phenomena mentioned in section 2.

Proposition 6 *The polynomial evaluated on* $[S^2] \in H_2(M;\mathbf{Z})$ *equals:*

$$\Phi_M([S^2]^{2k_0-2}) = (k_0)^{2k_0-2}.$$

One shows first that the elements in $\mathcal{M}(m_j, k_j)$ are described by elements of a Zariski open set of:

$$\mathbf{P}(H^1(S;\mathcal{O}(-2k_0)) = \mathbf{C}P^{2k_0-2}$$

defining extensions up to scalars of $\mathcal{O}(k_0)$ by $\mathcal{O}(-k_0)$ isomorphic to $\mathcal{O}(m)\oplus\mathcal{O}(-m)$ with $m < m_0$. Then one proceeds to identify $\mathcal{L}_{S^2}$ as the line bundle $\mathcal{O}(k_0)$ on $\mathbf{C}P^{2k_0-2}$. This establishes the result.

Finally we shall look at the case that both charges and masses are not zero. We shall identify the generator of $H_1(M, \partial M;\mathbf{Z})$ with a curve $\lambda = \{x\} \times \mathbf{R}$, for $x \in S^2$. The moduli space has dimension $4k_0 + 4k_\infty - 2$.

Proposition 7 *We have:*

$$\Phi_{M,m_j,k_j}([S^2]^{2k_0+2k_\infty-1}) = 0.$$

If $k_\infty \neq 0$ *then*

$$\Phi_{M,m_j,k_j}([S^2]^{2k_0+2k_\infty-4}[\lambda]^3) \neq 0$$

and if it is zero then

$$\Phi_{M,m_j,k_j}([S^2]^{2k_0-2}[\lambda]) \neq 0.$$

The vanishing observed in proposition 7 is not likely to be a general phenomenon.

Notice that the condition for a monopole to lie in the zero set of the section of $\mathcal{L}_\lambda$ can be formulated differently saying that the monopole must have the geodesic λ in its spectral curve (see Hitchin [H]). This leads one to:

Conjecture 8 *Let* γ *be a curve connecting two ends,* S_1 *and* S_2 *say, of a three manifold* M*, and denote the corresponding class in* $H_1(M, \partial M;\mathbf{Z})$ *by* λ*. Assume that* M *and* k_j, m_j *are such that the invariants are well defined. Then the coefficient of any term of* Φ_{M,m_j,k_j} *containing a power of* λ *bigger than* $k_1 + k_2 + 1$ *is 0.*

We end by showing how the connected sum conjectures could lead to big simplifications in understanding the invariants:

Conjecture 9 *Let* M *be a compact three manifold. Let* n *equal 0 when the real homology of* M *is non-zero and the order of* $H_1(M;\mathbf{Z})$ *otherwise. Then:*

$$\Phi_{M-2points}(k_j, m_j) = n\Phi_{S^2\times\mathbf{R}}(k_j, m_j).$$

Notice that $S^2 \times \mathbf{R}$ equals S^3 minus two points. This makes M minus two points a connected sum as in Conjecture 5.

5 References

[AM] Abkulut S. and McCarthy J.,*Cassons Invariant for oriented homology 3-Spheres - an Exposition*, Mathematical Notes Vol 36, PUP, 1990.
[B1] Braam P.,*Polynomial Invariants for 3-Manifolds and the Conformal Geometry of Magnetic Monopoles*, preprint.
[B2] Braam P.,*Magnetic Monopoles on Three Manifolds*, J.Diff. Geom., 30(1989), 425-464.
[DK] Donaldson S.K. and Kronheimer P.B.,*The Geometry of 4-Manifolds*, Oxford University Press, 1990.
[Fl1] Floer A., *Monopoles on Asymptotically Euclidean 3-Manifolds*, Bull. Ams. Vol.16 (1987), pp 125-127.
[Fl2] Floer A., *Instanton Homology, Surgery and Knots*, in *Geometry of Low dimensional Manifolds:1*, Eds. S.K.Donaldson and C.B.Thomas, LMS Lecture Note Series 150, CUP, 1990.
[H] Hitchin N., *Monopoles and Geodesics*, Comm. Math. Phys. 83(1982),579-602.
[JT] Jaffe A. and Taubes C., *Vortices and Monopoles*, Birhauser, Boston 1980.

5

Kawamata-Viehweg Vanishing Theorem for Compact Kähler Manifolds

ICHIRO ENOKI

Department of Mathematics, College of General Education, Osaka University, Osaka, Japan

0. Introduction

Related to the classification theory of algebraic varieties, cohomology groups of semi-positive or nef line bundles are of interest to many authors. Kawamata [**Ka**] and Viehweg [**V**] generalized the Kodaira vanishing theorem to nef line bundles over projective manifolds. Kollár [**Kol**] showed that the multiplication by a holomorphic section of a line bundle over a projective algebraic manifold induces injective homomorphisms between cohomology groups of the line bundle provided that certain tensor power of this line bundle is generated by global sections.

In this paper we shall extend these theorems, the vanishing theorem and the injectivity theorem, to compact Kähler manifolds. Moreover we shall give a theorem of another type, a strong-Lefschetz-type theorem. To give a precise statement of each result, let M be a compact Kähler manifold of dimension n with Kähler form ω. Let K_M denote the canonical line bundle of M. Let F be a line bundle over M. We call F *semi-positive* if the real first Chern class $c_1(F)_{\mathbb{R}} \in H^2(M, \mathbb{R})$ of F can be represented by a real d-closed $(1,1)$-form γ which is positive semi-definite everywhere on M; F is *nef* if, for any $\varepsilon > 0$, the first real Chern class $c_1(F)_{\mathbb{R}} \in H^2(M, \mathbb{R})$ of F can be represented by a real d-closed $(1,1)$-form γ_ε such that $\varepsilon\omega + \gamma_\varepsilon$ is positive definite everywhere on M. Clearly the semi-positivity implies the nef property. For a nef line bundle F over M, the *numerical Kodaira dimension* $\nu(F)$ of F is defined by

$$\nu(F) := \max\{k \mid c_1(F)_{\mathbb{R}}^k \neq 0 \text{ in } H^{2k}(M, \mathbb{R})\}.$$

Our vanishing theorem and injective theorem are the following:

Theorem 0.1. *Let F be a nef line bundle over a compact n-dimensional Kähler manifold M. Then $H^q(M, K_M + F) = 0$ for $q > n - \nu(F)$.*

Theorem 0.2. *Let F be a semi-positive line bundle over a compact Kähler manifold M. Suppose $F^{\otimes k}$, $k > 0$, admits a non-zero global holomorphic section s. Then the homomorphism*

$$\bullet \otimes s\colon H^q(M, K_M + mF) \to H^q(M, K_M + (m+k)F)$$

induced by the tensor product with s is injective for any $m > 0$ and q.

Note here that Maehara obtained independently the above generalization of Kollár's injectivity theorem.

Finally we shall state the strong-Lefschetz-type theorem. The classical strong Lefschetz theorem says that the multiplication with ω^q induces an isomorphism $L^q\colon H^{n-q}(M,\mathbb{C}) \to H^{n+q}(M,\mathbb{C})$; in particular, $L^q\colon H^0(M, \Omega_M^{n-q}) \to H^q(M, K_M)$ is an isomorphism by the Hodge decomposition. We consider the following more general situation: $H^0(M, \Omega_M^{n-q}(F)) \to H^q(M, K_M + F)$.

Theorem 0.3. *Let M be a compact Kähler manifold of dimension n with Kähler form ω. Let F be a semipositive line bundle over M. Then the homomorphism*

$$L^q\colon H^0(M, \Omega_M^{n-q}(F)) \to H^q(M, K_M + F)$$

induced by the multiplication with ω^q is surjective for every q.

We hope that the assumption that F is semipositive in Theorems 0.2 and 0.3 can be weakend. In this paper we shall restrict ourselves to the semipositive case for Theorems 0.2 and 0.3; only for the vanishing theorem we consider nef line bundles.

Acknowledgment. The author would like to thank Prof. Takegoshi who pointed out a gap in the first version of this note.

1. Kähler identities

Fixing our notations, we recall the Kähler identities on line bundles (cf. [**Ko**] or [**S**]). Let M be a compact Kähler manifold of dimension n with Kähler form ω. Let F be a line bundle over M with hermitian metric h. Let $A^{p,q}(F)$ be the space of F-valued smooth (p,q)-forms on M. Let $*: A^{p,q}(F) \to A^{n-q,n-p}(F^*)$ be the anti-complex linear Hodge star operator relative to the Kähler form ω and the hermitian metric h, where F^* is the dual bundle of F. Let $(\ , \ )$ be the L^2-inner product on $A^{p,q}(F)$ relative to h and ω. Let $L: A^{p,q}(F) \to A^{p+1,q+1}(F)$ be the homomorphism defined by the multiplication by the Kähler form ω. Let Λ be the formal adjoint of L. More generally, for any form α on M we define an operator $\ell(\alpha)$ by $\ell(\alpha)\xi = \alpha \wedge \xi$; and let $\lambda(\alpha)$ denote the formal adjoint of $\ell(\alpha)$.

Let ∇_h denote the hermitian connection of F relative to the hermitian metric h. Let ∇'_h be the $(1,0)$-component of ∇_h. (The $(0,1)$-component is the $\bar\partial$.) Moreover ∇'_h extends to the operator $\partial_h: A^{p,q}(F) \to A^{p+1,q}(F)$. We simply write ∂ for ∂_h if there is no danger of confusion. The curvature form γ of (F,h) is given by $\ell(\gamma) = \partial_h\bar\partial + \bar\partial\partial_h$. Let ∂_h^* and $\bar\partial^*$ denote, respectively, the formal adjoint operators of ∂_h and $\bar\partial$. The $\bar\partial$ Laplacian $\Box_{\bar\partial} := \bar\partial\bar\partial^* + \bar\partial^*\bar\partial$ can be written as

$$\Box_{\bar\partial} = \partial_h\partial_h^* + \partial_h^*\partial_h + \sqrt{-1}(\ell(\gamma)\Lambda - \Lambda\ell(\gamma)). \tag{1.1}$$

This formula can be derived from the following fundamental identities:

$$\begin{aligned} \partial_h^* L - L\partial_h^* &= -\sqrt{-1}\bar\partial, \quad \bar\partial^* L - L\bar\partial^* = \sqrt{-1}\partial_h \\ \partial_h\Lambda - \Lambda\partial_h &= -\sqrt{-1}\bar\partial^*, \quad \bar\partial\Lambda - \Lambda\bar\partial = \sqrt{-1}\partial_h^*. \end{aligned} \tag{1.2}$$

2. Lefschetz-type theorem and injectivity theorem

The following proposition implies Theorems 0.2 and 0.3.

Proposition 2.1. *Let (F, h) be an hermitian line bundle over a compact Kähler manifold of dimension n. Suppose (F, h) is semipositive. Then an F-valued (n, q)-form ξ is harmonic, $\Box_{\bar\partial}\xi = 0$, if and only if*

$$\bar\partial \Lambda^q \xi = 0 \quad \textit{and} \quad \ell(\gamma)\Lambda\xi = 0.$$

Proof. By (1.1)

$$\|\bar\partial\xi\|^2 + \|\bar\partial^*\xi\|^2 = (\Box_{\bar\partial}\xi, \xi) = \|\partial_h^*\xi\|^2 + (\sqrt{-1}\ell(\gamma)\Lambda\xi, \xi), \tag{2.2}$$

Note that $\partial_h\xi = 0$ since ξ is of type (n, q). Let $\eta = \frac{1}{q!}\Lambda^q\xi$ so that $\xi = \frac{1}{q!}L^q\eta$. Then, by (1.2)

$$q!\partial_h^*\xi = -\sqrt{-1}qL^{q-1}\bar\partial\eta - \sqrt{-1}L^q\Lambda\bar\partial\eta. \tag{2.3}$$

Suppose $\bar\partial\eta = 0$ and $\ell(\gamma)\Lambda\xi = 0$. Then $\partial_h^*\xi = 0$ by (2.3). In view of (2.2) it follows that $\bar\partial\xi = 0$ and $\bar\partial^*\xi = 0$, i.e., ξ is harmonic. Conversely, suppose that ξ is harmonic. Then $q!\bar\partial\xi = L^q\bar\partial\eta = 0$ by (1.2). This means that $\bar\partial\eta$ is primitive and hence $\Lambda\bar\partial\eta = 0$. It follows by (2.3) that $\|\partial_h^*\xi\| = \|\bar\partial\eta\|$. On the other hand the point-wise linear endmorphism $\sqrt{-1}\ell(\gamma)\Lambda$ of $\bigwedge^{n,q} TM \otimes F$ is self adjoint and all eigenvalues are nonnegative since γ is semipositive. In view of (2.2) it follows that $\bar\partial\eta = 0$ and $\ell(\gamma)\Lambda\xi = 0$. ∎

Proof of Theorem 0.2. Let $s \in H^0(M, F^{\otimes k})$ and let ξ be an $F^{\otimes m}$-valued harmonic (n, q)-form. It suffices to show that $s \otimes \xi$ is an $F^{\otimes(m+k)}$-valued harmonic (n, q)-form. Fix an hermitian metric h of F such that its curvature form γ is semipositive; we equip the induced hermitian metrics on $F^{\otimes m}$ and $F^{\otimes(m+k)}$ so that the curvature forms are $m\gamma$ and $(m + k)\gamma$ respectively. Applying Proposition 2.1to ξ we obtain that $\bar{\partial}\Lambda^q\xi = 0$ and $m \cdot \ell(\gamma)\Lambda\xi = 0$. The first condition implies that $\bar{\partial}\Lambda^q(s \otimes \xi) = \bar{\partial}(s \otimes \Lambda^q\xi) = 0$, since $\bar{\partial}s = 0$. Since s is of type $(0,0)$, the second condition implies that $\ell((m+k)\gamma)\Lambda(s \otimes \xi) = (m+k)s \otimes \ell(\gamma)\Lambda\xi) = 0$. Applying Proposition 2.1now to $F^{\otimes(m+k)}$, we obtain therefore that $s \otimes \xi$ is harmonic.

3. Vanishing theorem

The Nakano inequality says that $\sqrt{-1}((\ell(\gamma)\Lambda - \Lambda\ell(\gamma))\xi, \xi) \leq 0$ if ξ is harmonic. This can be generalized as follows.

Proposition 3.1. *If an F-valued (n,q)-form ξ is harmonic, then*

$$\sqrt{-1}(\ell(\gamma + \partial\bar{\partial}f)\Lambda\xi, e^f\xi) \leq 0$$

for any smooth real function f on M.

Proof. We show first

$$\partial_h^*(e^f\xi) = e^f\partial_h^*\xi - e^f\lambda(\partial f)\xi. \tag{3.2}$$

In fact, for any F-valued $(n-1,q)$-form φ we have

$$\begin{aligned}(\partial_h^*(e^f\xi), \varphi) &= (\xi, e^f\partial\varphi) \\ &= (\xi, \partial(e^f\varphi) - (\partial e^f) \wedge \xi) \\ &= (e^f\partial_h^*\xi - e^f\lambda(\partial f)\xi, \varphi).\end{aligned}$$

Since $\bar{\partial}^*\xi = 0$, we have $\partial_h\Lambda\xi = 0$, using (1.2). Thus $(\partial\bar{\partial}f) \wedge \Lambda\xi = \partial_h(\bar{\partial}f \wedge \Lambda\xi)$. Therefore by (3.2)

$$\begin{aligned}(\partial\bar{\partial}f \wedge \Lambda\xi, e^f\xi) &= (\partial_h(\bar{\partial}f \wedge \Lambda\xi), e^f\xi) \\ &= (\bar{\partial}f \wedge \Lambda\xi, \partial_h^*(e^f\xi)) \\ &= (\bar{\partial}f \wedge \Lambda\xi,, e^f\partial_h^*\xi) - (\bar{\partial}f \wedge \Lambda\xi, e^f\lambda(\partial f)\xi).\end{aligned} \tag{3.3}$$

By (1.1) it follows from $(\square_{\bar\partial}\xi, e^f\xi) = 0$ that

$$\begin{aligned}\sqrt{-1}(\ell(\gamma)\Lambda\xi, e^f\xi) &= -(\partial_h^*\xi, \partial_h^*(e^f\xi)) \\ &= -(\partial_h^*\xi, e^f\partial_h^*\xi) + (\partial_h^*\xi, e^f\lambda(\partial f)\xi)\end{aligned}$$

Let $\eta = \frac{1}{q!}\Lambda^q\xi$ so that $\xi = \frac{1}{q!}L^q\eta$. Then it follows from $\bar\partial\xi = 0$ that $\bar\partial\eta$ is primitive. Using(1.2), we obtain

$$(\partial_h^*\xi, e^f\partial_h^*\xi) = (\bar\partial\eta, e^f\bar\partial\eta).$$

Making also use of the identities,

$$\lambda(\partial f)L - L\lambda(\partial f) = \sqrt{-1}\ell(\bar\partial f),$$

we can derive

$$\begin{aligned}(\partial_h^*\xi, e^f\lambda(\partial f)\xi) &= -(\bar\partial\eta, e^f\bar\partial f\wedge\eta), \\ (\sqrt{-1}\bar\partial f\wedge\Lambda\xi, e^f\partial_h^*\xi) &= -(\bar\partial f\wedge\eta, e^f\bar\partial\eta), \\ (\sqrt{-1}\bar\partial f\wedge\Lambda\xi, e^f\lambda(\partial f)\xi)) &= ((\bar\partial f\wedge\eta)_0, e^f(\bar\partial f\wedge\eta)_0),\end{aligned}$$

where $(\bar\partial f\wedge\eta)_0$ denotes the primitive part of $\bar\partial f\wedge\eta$. Combining these together we conclude

$$\sqrt{-1}(\ell(\gamma+\partial\bar\partial f)\Lambda\xi, e^f\xi) = -\|e^{\frac{f}{2}}(\bar\partial\eta + (\bar\partial f\wedge\eta)_0)\|^2 \le 0. \quad \blacksquare$$

We will use the following easy lemmas.

Lemma 3.4. *Let M be a compact Riemannian manifold with the Laplacian $\triangle$. Let $\{f_j\}$ be a sequence of smooth functions on M with $\int_M f_j = 0$. Suppose $\lim_{j\to\infty} \triangle f_j = 0$ as current on M. Then $\lim_{j\to\infty} f_j = 0$ as current on M.*

Proof. For a smooth test function φ, set $\varphi = c + \varphi_0$ with $c = \int_M \varphi$. Note that $\langle f_j, \varphi\rangle = \langle f_j, \varphi_0\rangle$ since $\int_M f_j = 0$. Let G be the Green operator for the Laplacian $\triangle$. Then

$$\langle f_j, \varphi\rangle = \langle f_j, \triangle G\varphi_0\rangle = \langle \triangle f_j, G\varphi_0\rangle \to 0 \quad \text{as } j \to 0. \blacksquare$$

Lemma 3.5. *Let $\{f_j\}$ be a sequence of non-positive smooth functions on M. Suppose that $\{f_j\}$ converges as current. Then $\lim_{j\to\infty} \int_U e^{f_j} \neq 0$ for any open subset U of M.*

Proof. Suppose $\lim_{j\to\infty} \int_U e^{f_j} = 0$ for some open subset U. Take a relatively compact open subset $U_0 \subset U$ and a smooth non-negative function φ supported on U with $\varphi\,|_{U_0}= 1$. Set $\varepsilon_j := \int_{U_0} e^{f_j}$ and $V_j := \{p \in U_0 \mid e^{f_j(p)} < 2\varepsilon_j\}$. Then $\lim_{j\to\infty} \varepsilon_j = 0$ and $2Vol(V_j) > Vol(U_0)$. Therefore

$$\int_M \varphi f_j \leq Vol(U_j) \log 2\varepsilon_j \to -\infty \quad \text{as } j \to \infty,$$

which contradicts to the convergence of $\{f_j\}$ as current. $\blacksquare$

Proof of Theorem 0.1. We fix an hermitian metric h of F and a Kähler form ω of M. Let ξ be an F-valued harmonic (n,q)-form relative to h and ω. Let $\nu = \nu(F)$ and $n = \dim M$. Supposing $q > n - \nu$ and $\xi \neq 0$, we shall derive a contradiction. Let

$$p(\varepsilon) := \int_M (\varepsilon\omega + \sqrt{-1}\gamma)^n / \int_M \omega^n,$$

where γ is the curvature form of (F, h). Since F is nef, for each $\varepsilon > 0$ there is a real smooth function ψ_ε such that $\varepsilon\omega + \sqrt{-1}\gamma + \sqrt{-1}\partial\bar{\partial}\psi_\varepsilon$ is a Kähler form on M. According to the solution of the Calabi conjecture by Yau [**Y**], for each $\varepsilon > 0$ there is a real smooth function φ_ε on M such that $\varepsilon\omega + \sqrt{-1}\gamma + \sqrt{-1}\partial\bar{\partial}(\psi_\varepsilon + \varphi_\varepsilon)$ is a Kähler form on M with

$$(\varepsilon\omega + \sqrt{-1}\gamma + \sqrt{-1}\partial\bar{\partial}(\psi_\varepsilon + \varphi_\varepsilon))^n = p(\varepsilon)\omega^n.$$

Let $f_\varepsilon = \psi_\varepsilon + \varphi_\varepsilon$ and set $\rho_\varepsilon = \varepsilon\omega + \sqrt{-1}(\gamma + \partial\bar{\partial} f_\varepsilon)$. Since $\int_M \rho_\varepsilon \wedge \omega^{n-1}$ is bounded as $\varepsilon \to 0$, there is a sequence $\{\varepsilon_j\}$ with $\lim_{j\to\infty} \varepsilon_j = 0$ such that $\lim_{j\to\infty} \rho_{\varepsilon_j}$ exists as current on M. Then $\{f_{\varepsilon_j}\}$ converges as current under the normalization $\int_M f_{\varepsilon_j} = 0$ by Lemma 3.4. Moreover, since $\sqrt{-1}\partial\bar{\partial} f_{\varepsilon_j} > -c\omega$ with constant c independent of ε_j, we have $\sup_M f_{\varepsilon_j} \leq C$ with constant C independent of ε_j. To simplify notations we set $\varepsilon = \varepsilon_j$ and hence $f_\varepsilon = f_{\varepsilon_j}$, $\rho_\varepsilon = \rho_{\varepsilon_j}$. Moreover, when we let $j \to \infty$ (and $\varepsilon_j \to 0$), we simply write $\varepsilon \to 0$.

Choose a local unitary frame $(\theta^1, \ldots, \theta^n)$ of the holomorphic cotangent bundle of M so that ρ_ε is diagonal with respect to this frame: $\rho_\varepsilon = \sum_i \lambda_i \sqrt{-1}\theta^i \wedge \bar{\theta}^i$. Thus $\prod_{i=1}^n \lambda_i = p(\varepsilon)$. Note that $p(\varepsilon)$ is a polynomial in ε of degree n. Let $p(\varepsilon) = \sum_{j=0}^n a_j \varepsilon^j$. Since F is nef, we have $a_j > 0$ for $j \geq n - \nu$ and $a_k = 0$ for $k < n - \nu$ by the definition of the numerical Kodaira dimension ν. Consequently

$$a_{n-\nu}\varepsilon^{n-\nu} \leq \prod_i \lambda_i, \tag{$*$}$$

where $a_{n-\nu} > 0$ is independent of ε.

Consider a non-negative continuous function $u_\varepsilon := \langle \ell(\rho_\varepsilon)\Lambda\xi, \xi\rangle / \langle \xi, \xi\rangle$ and let $U(\varepsilon) := \{x \in M \mid u_\varepsilon(x) < 2\varepsilon\}$. We claim that $\int_{U(\varepsilon)} \omega^n \geq C_0$ for some constant $C_0 > 0$ independent of ε. Suppose the contrary: $\int_{U(\varepsilon)} \omega^n \to 0$ as $\varepsilon \to 0$. By

Proposition 3.1, we have $\int_M u_\varepsilon dv_\varepsilon \leq \varepsilon q \int_M dv_\varepsilon$, where $dv_\varepsilon := \langle \xi, \xi \rangle e^{f_\varepsilon} \omega^n$. (Since ξ is of type (n, q), we have $\omega \wedge \Lambda \xi = q\xi$.) Hence $\int_{U(\varepsilon)} dv_\varepsilon \geq \frac{q}{2} \int_M dv_\varepsilon$. Then, noting that $\sup_M \langle \xi, \xi \rangle e^{f_\varepsilon}$ is bounded as $\varepsilon \to 0$, we have

$$\frac{q}{2} \int_M \langle \xi, \xi \rangle e^{f_\varepsilon} \omega^n \leq \int_{U(\varepsilon)} \langle \xi, \xi \rangle e^{f_\varepsilon} \omega^n \leq K \int_{U(\varepsilon)} \omega^n \to 0,$$

where K is a constant independent of ε such that $\langle \xi, \xi \rangle e^{f_\varepsilon} \leq K$ on M. On the other hand, since ξ is harmonic, ξ is not identically zero on any open subset by Aronszajn's unique continuation theorem for solutions of elliptic equations. Hence there is an open subset U on which ξ never vanishes. Then the above inequality implies that $\int_U e^{f_\varepsilon} \to 0$ as $\varepsilon \to 0$. Since $\sup_M f_\varepsilon$ is bounded as $\varepsilon \to 0$, this contradicts to Lemma 3.5.

$n \int_{U(\varepsilon)} \sum_{\beta=1}^n \lambda_\beta \omega^n (\leq \int_M \rho_\varepsilon \wedge \omega^{n-1})$ is bounded as $\varepsilon \to 0$, while $\int_{U(\varepsilon)} \omega^n \geq C_0$. Therefore there is a constant C independent of ε such that $\sum_{\beta=1}^n \lambda_\beta \leq C$ (and hence $\lambda_\beta \leq C$) on an open subset $V(\varepsilon)$ of $U(\varepsilon)$.

For an ordered set $A = \{i_1, \ldots, i_q\}$ we write $\theta^A = \theta^{i_1} \wedge \cdots \wedge \theta^{i_q}$. Letting $N = \{1, \ldots, n\}$, we put $\xi = \sum_A \xi_A \theta^N \wedge \bar{\theta}^A$. Then we have

$$(**) \qquad \ell(\rho_\varepsilon) \Lambda \xi = \sum_A \Big(\sum_{\alpha \in A} \lambda_\alpha\Big) \xi_A \theta^N \wedge \bar{\theta}^A.$$

By Aronszajn's theorem, ξ is not identically zero on $V(\varepsilon)$. Shrinking $V(\varepsilon)$ if necessary, we may assume there is an index set A_0 such that ξ_{A_0} never vanishes on $V(\varepsilon)$ and $|\xi_A|_h \leq |\xi_{A_0}|_h$ on $V(\varepsilon)$ for any A, where $|\ |_h$ denotes the point-wise norm relative to h. Then $\langle \xi, \xi \rangle \leq \binom{n}{q} |\xi_{A_0}|_h^2$ and $\sum_{i \in A_0} \lambda_i |\xi_{A_0}|_h^2 \leq \langle \ell(\rho_\varepsilon) \Lambda \xi, \xi \rangle$ in view of $(**)$. Thus we obtain by the inequality $u_\varepsilon \leq 2\varepsilon$ that $\lambda_\alpha \leq C\varepsilon$ on $V(\varepsilon)$ for $\alpha \in A_0$, letting C be larger independently to ε if necessary. Consequently $\prod_{\alpha=1}^n \lambda_\alpha \leq C^n \varepsilon^q$. By $(*)$ it follows $a_{n-\nu} \varepsilon^{n-\nu} \leq C^n \varepsilon^q$ as $\varepsilon \to 0$. This is a contradiction because $n - \nu < q$. ∎

References

[**Ka**] Kawamata, Y., *A generalization of Kodaira-Ramanujam's vanishing theorem*, Math. Ann. **261** (1982), 43-46.

[**Ko**] Kobayashi, S., "Differential Geometry of Complex Vector Bundles," Iwanami Shoten, Tokyo, and Princeton Univ. Press, Princeton, 1987.

[**Kol**] Kollár, J., *Higher direct images of dualizing sheaves I*, Ann. of Math. **123** (1986), 11-42.

[**S**] Siu, Y.-T., *Complex-analyticity of harmanic maps, vanishing theorems*, J. Diff. Geometry **17** (1982), 55-138.

[**V**] Viehweg, E., *Vanishing theorems*, J. Reine Angew. Math. **335** (1982), 1-8.

[**Y**] Yau, S.-T., *On the Ricci curvature of a compact Kähler manifold and the complex Monge-Ampère equation, I*, Comm. Pure Appl. Math. **31** (1978), 339-411.

6

Morse Theory and Thom-Gysin Exact Sequence

M. Furuta University of Tokyo, Tokyo, Japan

In this article we explain how to understand the Thom-Gysin exact sequence for a sphere bundle in context of Morse theory (section 1). A similar construction is possible for any cohomology class which is not necessarily of form of the Euler class of a sphere bundle (section 2). In section 3 an application of this construction to the Floer cohomology is described, which is a joint work with K. Fukaya and H. Ohta [FFO]. In fact the construction was inspired by a work of Fukaya [Fk] which gives a generalization of the Floer homology for general oriented 3-manifolds.

§1 Morse theory and S^1-bundle

In this section using Morse theory we give a description of the Thom-Gysin exact sequence associated to an S^1-bundle.

1-1 Morse theory.

We begin with review of the Morse theory. The Morse theory can be formulated as a description of cohomology groups of a closed manifold by counting numbers of gradient flow lines between critical points of a Morse function on the manifold.

Suppose f is a Morse function on a (finite dimensional) closed manifold M. Take a generic Riemannian metric on M and let $V = -\operatorname{grad} f$. For a critical point p, let S_p (U_p) be the stable (unstable) manifold of p defined as the set of points which tend to p along flow lines generated by V ($-V$). Then S_p (U_p) has structure of smooth manifold. we have the decomposition $(TM)_p = (TS_p)_p \oplus (TU_p)_p$. There is a diffeomorphism between S_p (U_p) and $(TS_p)_p$ ($(TU_p)_p$). This diffeomorphism can be constructed by using the flow lines. The Morse index $\lambda(p)$ of p is the dimension of $(TU_p)_p$. Notice that to fix one of the two orientations of $(TU_p)_p$ is equivalent to fix one of the two generators of $C(p) = H_c^{\lambda(p)}((TU_p)_p, \mathbf{Z}) \cong \mathbf{Z}$. This choice is furthermore equivalent to fix an orientation of U_p, or a coorientation of S_p, i.e., an orientation of the normal bundle of S_p.

Let p and q be critical points with $\lambda(p) \leq \lambda(q)$, and $\mathcal{M}(p,q)$ the set of points which tend to p along V and to q along $-V$, i.e., $\mathcal{M}(p,q) = S_p \cap U_q$. When the metric of M is generic, all the intersection of the form $S_p \cap U_q$ are transverse, and dimension counting implies that the dimension of $\mathcal{M}(p,q)$ is equal to $\lambda(q) - \lambda(p)$. A choice of the generators of $C(p)$ and $C(q)$ determines an orientation of $\mathcal{M}(p,q)$ through the orientation of U_q and the coorientation of S_p. When $\lambda(q)$ is strictly larger than $\lambda(p)$, the flow induces a free $\mathbf{R}$-action on $\mathcal{M}(p,q)$, and the dimension of $\overline{\mathcal{M}}(p,q) = \mathcal{M}(p,q)/\mathbf{R}$ is $\lambda(q) - \lambda(p) - 1$. If c is an real number which satisfies $f(p) < c < f(q)$, then $\overline{\mathcal{M}}(p,q)$ can be identified with $\mathcal{M}(p,q) \cap f^{-1}(c)$. If $f(q) \leq f(p)$ and $q \neq p$, then $M(p,q) = \emptyset$.

Let C^λ be the direct sum of $C(p)$ for all critical points p with $\lambda(p) = \lambda$. Then one can define a linear map $d : C^\lambda \longrightarrow C^{\lambda+1}$ as follows.

Suppose $\lambda(p) = \lambda$ and $\lambda(q) = \lambda + 1$. Then $\mathcal{M}(p,q)$ is one-dimensional and oriented once the generators of $C(p)$ and $C(q)$ are fixed. Moreover it can be shown that $\mathcal{M}(p,q)$ consists of the union of finitely many flow lines, i.e., $\overline{\mathcal{M}}(p,q)$ consists of finitely many points since we can use the following lemma (for the case $k = 1$) to see compactness of $\mathcal{M}(p,q) \cap f^{-1}(c)$ for a regular value c of f, which is identified with $\overline{\mathcal{M}}(p,q)$ if $f(p) < c < f(q)$.

LEMMA 1.1. *For a non-negative integer k, the union of all $\mathcal{M}(r,r')$ with $\lambda(r) - \lambda(r') \leq k$ is a compact subset of M. More precisely, for two critical points p and q with $\lambda(q) - \lambda(p) = k$, the union of all $\mathcal{M}(r,r')$ with the following condition is compact; there is a finite sequence of critical points $r_0 = p, r_1, \cdots, r_l = q$, starting from p and ending at q, such that $\mathcal{M}(r_i, r_{i+1})$ is not empty for each i and that $r = r_i$ and $r' = r_{i+1}$ for some i.* ∎

For a moment we write p and q for the fixed generators of $C(p)$ and $C(q)$. The map d is defined by

$$dp = \sum_{\lambda(q)=\lambda+1} (\#\overline{\mathcal{M}}(p,q))q,$$

where $\#\overline{\mathcal{M}}(p,q)$ is the number of the points of $\overline{\mathcal{M}}(p,q)$ counted with sign. The following theorem is now well known as a formulation of the Morse theory.

THEOREM 1.2.

(1) $dd = 0$.

(2) The cohomology of the complex (C^*, d) is canonically isomorphic to $H^*(M, \mathbf{Z})$.

Remark Similarly the $\mathbf{Z}_2$-coeficient cohomology of M is given by the cohomology of $(C^* \otimes \mathbf{Z}_2, d)$. In this case we do not have to consider the various orientations which is necessary in the definition of (C^*, d).

1-2 S^1-bundle.

Let e be a $\mathbf{Z}$-cohomology class of degree 2. Then there is a principal S^1-bundle $\pi : P \longrightarrow M$ be over M with Euler class $e \in H^2(M, \mathbf{Z})$. In this sense P could be seen as a geometric realization of e. The Thom-Gysin exact sequence associated to P is the long exact sequence:

$$\longrightarrow H^{i-2}(M,\mathbf{Z}) \xrightarrow{e\cdot} H^i(M,\mathbf{Z}) \xrightarrow{\pi^*} H^i(P,\mathbf{Z}) \xrightarrow{\pi_*} H^{i-1}(M,\mathbf{Z}) \longrightarrow$$

where $e\cdot$ is the cup product with e, π^* the map naturally induced by π, and π_* the integration along fibre for π. We shall give a description of this sequence, in particular that of $H^i(P, \mathbf{Z})$, in terms of Morse theory.

Let $L = P \times_{S^1} \mathbf{C}$ be the associated complex line bundle with P. Take a section s of $L \longrightarrow M$ transverse to the zero section. The zero set $s^{-1}(0)$ is a smooth cooriented submanifold of codimension 2 whose Poincaré dual is e. We moreover assume that (1) $s^{-1}(0)$ does not contain any critical point, (2) $s^{-1}(0)$ does not intersect with any $\mathcal{M}(p,q)$ with $\lambda(p) + 1 = \lambda(q)$, and (3) $s^{-1}(0)$ intersects transversely with any $\mathcal{M}(p,q)$ with $\lambda(p) + 2 = \lambda(q)$. A generic s satisfies this assumption. Using f and s we define a linear map $\tilde{e}(s) : C^\lambda \longrightarrow C^{\lambda+2}$ as follows. Fix a generator of $C(p)$ for each critical point p and denote it also by p as in a previous argument. For critical points p and q with $\lambda(p) = \lambda$ and $\lambda(q) = \lambda + 2$, the dimension of the intersection $\mathcal{M}(p,q) \cap s^{-1}(0)$ is zero, as dim $\mathcal{M}(p,q) = 2$ and codim $s^{-1}(0) = 2$. lemma 1.1 and the assumption for s imply compactness of $\mathcal{M}(p,q) \cap s^{-1}(0)$, hence this intersection is a finite set. Let $\#(\mathcal{M}(p,q) \cap s^{-1}(0))$ be the number of the points counted with sign. Now the definition of $\tilde{e}(s)$ is

$$\tilde{e}(s)p = \sum_{\lambda(q)=\lambda+2} (\#(\mathcal{M}(p,q) \cap s^{-1}(0)))q.$$

THEOREM 1.3. $d\tilde{e}(s) \pm \tilde{e}(s)d = 0$.

Here the sign depends on the convention of the orientation in the above construction. We take a convention for which the sign is +. Then theorems 1.2 (1) and 1.3 can be summerized as

$$\begin{pmatrix} d & \tilde{e}(s) \\ 0 & d \end{pmatrix}^2 = 0.$$

We denote by $\tilde{d}$ the above matrix whose square is zero. We can think of $\tilde{d}$ as a coboundary map of a differencial complex. We denote by $C^* \oplus C^{*-1}$ the graded $\mathbf{Z}$-module

$$(\bigoplus_\lambda C^\lambda) \oplus (\bigoplus_\lambda C^{\lambda-1}) = \bigoplus_\lambda (C^\lambda \oplus C^{\lambda-1}),$$

where the degree of $(C^\lambda \oplus C^{\lambda-1})$ is defined to be λ. Then $\tilde{d}$ gives a degree-one map on $C^* \oplus C^{*-1}$ with square zero.

THEOREM 1.4. *The cohomology of the differential complex* $(C^* \oplus C^{*-1}, \tilde{d})$ *is canonically isomorphic to* $H^*(P, \mathbf{Z})$.

Since $\tilde{d}$ is a upper triangle matrix, the submodule $C^* \oplus \{0\}$ is preserved by $\tilde{d}$, and we have the following short exact sequence of differential complexes:

$$0 \longrightarrow C^* \longrightarrow C^* \oplus C^{*-1} \longrightarrow C^{*-1} \longrightarrow 0,$$

where C^{*-k} is the differential complex (C^*, d) with degree shifted by $-k$. From theorems 1.2 (2) and 1.4, we can write the long exact sequence associated to this sequence as:

$$\longrightarrow H^\lambda(M, \mathbf{Z}) \longrightarrow H^\lambda(P, \mathbf{Z}) \longrightarrow H^{\lambda-1}(M, \mathbf{Z}) \longrightarrow .$$

THEOREM 1.5. *This exact sequence agrees with the Thom-Gysin exact sequence associated to* $P \longrightarrow M$.

In particular the maps in the long exact sequence does not depend on the choice of f, s nor the Riemannian metric on M.

Sketch of proofs of the theorems in this section shall be given in the next section, as proofs of theorems for more general sphere bundles.

§2 COHOMOLOGY CLASSES AND SPHERE BUNDLES

In this section we state an extension of the construction of the previous section.

2-1 Sphere bundle.

Let α be a $\mathbf{Z}$-cohomology class of degree n. If α is the Euler class of a sphere bundle $\pi : S \longrightarrow M$ be an S^{n-1}-bundle with structure group $SO(n)$, then the construction of section 1 can be extended as follows. The Thom-Gysin exact sequence associated to S is the long exact sequence:

$$\longrightarrow H^{i-n}(M,\mathbf{Z}) \xrightarrow{\alpha\cdot} H^i(M,\mathbf{Z}) \xrightarrow{\pi^*} H^i(S,\mathbf{Z}) \xrightarrow{\pi_*} H^{i-n+1}(M,\mathbf{Z}) \longrightarrow .$$

Let E be the real oriented vector bundle associated to S. Take a section s of $E \longrightarrow M$ transverse to the zero section. The zero set $s^{-1}(0)$ is a smooth cooriented submanifold of codimension n whose Poincaré dual is α. We moreover assume that (1) $s^{-1}(0)$ does not contain any critical point, (2) $s^{-1}(0)$ does not intersect with any $\mathcal{M}(p,q)$ with $\lambda(q) - \lambda(p) < n$, and (3) $s^{-1}(0)$ intersects transversely with any $\mathcal{M}(p,q)$ with $\lambda(q) - \lambda(p) = n$. A generic s satisfies this assumption as in section 1. Using f and s we define a linear map $\tilde{\alpha}(s) : C^\lambda \longrightarrow C^{\lambda+n}$ by

$$\tilde{\alpha}(s)p = \sum_{\lambda(q)=\lambda+n} (\#(\mathcal{M}(p,q) \cap s^{-1}(0)))q.$$

THEOREM 2.1. *We have $d\tilde{\alpha}(s) \pm \tilde{\alpha}(s)d = 0$, hence, with an appropriate sign convention,*

$$\begin{pmatrix} d & \tilde{\alpha}(s) \\ 0 & d \end{pmatrix}^2 = 0.$$

We denote by $\tilde{d}$ the above matrix whose square is zero. We can think of $\tilde{d}$ as a coboundary map of a differencial complex $C^* \oplus C^{*-n+1}$.

THEOREM 2.2. *The cohomology of the differential complex $(C^* \oplus C^{*-n+1}, \tilde{d})$ is canonically isomorphic to $H^*(S,\mathbf{Z})$.*

THEOREM 2.3. *The exact sequence associated to the short exact sequence*

$$0 \longrightarrow C^* \longrightarrow C^* \oplus C^{*-1} \longrightarrow C^{*-1} \longrightarrow 0$$

agrees with the Thom-Gysin exact sequence associated to $P \longrightarrow M$.

SKETCH OF PROOF OF THEOREMS 2.1 AND 2.2: Let $g : S \longrightarrow \mathbf{R}$ be the map defined by inner product with s at each fibre. If $s(x) = 0$, then $g = 0$ on the fibre $(S)_x$ at x. If $s(x) \neq 0$, then the restriction of g on $(S)_x$ has precisely two critical points. Fix a connection on $S \longrightarrow M$ so that along the finitely many gradient flows which are used to define $\tilde{d}$, the direction of s is parallel. Lift the gradient vector field $grad\, f$ to S. Let $grad_f\, g$ be the gradient vector field of g along fibres. Then $grad\, F + grad_f\, g$ is a *gradient-like* vector field for the function $\tilde{f} = \pi^* f + g$. For each critical point of f, there are exactly two critical points of $\tilde{f}$. Then the Morse complex for $(S, \tilde{f})$ is given by $(C^* \oplus C^{*-n+1}, \tilde{d})$. ∎

2-2 Cohomology classes. Because of the fact $S^1 = K(\mathbf{Z}, 2)$, there is a one to one correspondence between $H^2(M, \mathbf{Z})$ and the isomorphism classes of S^1-bundles over M. We can give a similar construction starting from any cohomology class $\alpha \in H^n(M, R)$, where R is a commutative ring with unit and $n \geq 2$. We first realise α as a primary obstruction class for a fibration over M to have a section.A standard way to do this is to use a continuous map A from M to the Eilenberg MacLane space $K(R, n)$ corresponding to α. Let $p : E(R, n) \longrightarrow K(R, n)$ be the fibration over $K(R, n)$ with fibre $K(R, n-1)$ which is defined to be the space of paths on $K(R, n)$ starting from a base point. The projection p is the map of taking the end point of a path and the fibre is $\Omega K(R, n) = K(R, n-1)$. Then, the pull back $A^*E(R, n)$ is the required fibration. This is a universal fibration with primary obstruction α in the following sense. For any CW-complex E and any map $f : E \longrightarrow M$ with $f^*\alpha = 0 \in H^n(E, R)$, there is a lift $E \longrightarrow E(R, n)$ which covers A, hence a fibre preserving map $E \longrightarrow A^*E(R, n)$. Let $f : E \longrightarrow M$ be a fibration with primary obstruction α.

LEMMA 2.4. *There is a section of $f : E \longrightarrow M$ over an open neighbourhood of the union of $\mathcal{M}(p, q)$ with $\lambda(q) - \lambda(p) < n$.*

Let $\mathcal{U}$ be the open set and s the section over $\mathcal{U}$. For p and q with $\lambda(q) - \lambda(p) = n$, the pair $(\mathcal{M}(p, q), \mathcal{M}(p, q) \cap \mathcal{U})$ has the n-dimensional fundamental class $[\mathcal{M}(p, q)] \in H_n(\mathcal{M}(p, q), \mathcal{M}(p, q) \cap \mathcal{U}; R)$, once an orientation of $\mathcal{M}(p, q)$ is fixed. Define the class $\alpha(p, q) \in H^n(\mathcal{M}(p, q), \mathcal{M}(p, q) \cap \mathcal{U}; R)$ to be the primary obstruction class for s to be extended over $\mathcal{M}(p, q)$ up to homotopy. Then we have a number $\alpha(p, q)[\mathcal{M}(p, q)] \in R$. Using f and s we define a linear map $\tilde{\alpha}(s) : C^\lambda \otimes R \longrightarrow C^{\lambda+n} \otimes R$ by

$$\tilde{\alpha}(s)p = \sum_{\lambda(q)=\lambda+n} \alpha(p, q)[\mathcal{M}(p, q)]q.$$

THEOREM 2.5.

(1) *With an appropriate sign convention, we have $\tilde{d}^2 = 0$ for*

$$\tilde{d} = \begin{pmatrix} d & \tilde{\alpha}(s) \\ 0 & d \end{pmatrix}.$$

(2) *The cohomology of the differential complex $((C^* \oplus C^{*-n+1}) \otimes R, \tilde{d})$ does not depend on the choice of f and s. It only depend on α.*

(3) *The exact sequence associated to the short exact sequence*

$$0 \longrightarrow C^* \otimes R \longrightarrow (C^* \oplus C^{*-1}) \otimes R \longrightarrow C^{*-1} \otimes R \longrightarrow 0$$

only depend on α.

If α is the Euler class of a sphere bundle, the cohomology group is canonically isomorphic to that of the total space of the sphere bundle (theorem 2.2). We can also identify the cohomology group for $\alpha \in H^{2m}(M, \mathbf{Q})$.

THEOREM 2.6. *For* $\alpha \in H^{2m}(M, \mathbf{Q})$,

$$H^i((C^* \oplus C^{*-2m+1}) \otimes Q, \tilde{d}) = H^i(A^* E(\mathbf{Q}, 2m), \mathbf{Q}).$$

SKETCH OF PROOF: Since the rational homotopy type of $BSO(2m)$ is the product of Eilenberg-MacLane spaces corresponding to the Euler class and the Pontrjagin classes, if we replace α by a multiple of α, the map $A : M \longrightarrow K(\mathbf{Q}, 2m)$ can be lifted to a map to $BSO(2m)$ such that the pull back of the Euler class is α. Then the theorem is reduced to theorem 2.2. ∎

§3 FLOER COHOMOLOGY AND A $\mathbf{Z}_2$-COHOMOLOGY CLASS.

3-1 Review of Floer cohomology.

Let Y be an oriented homology 3-sphere. Formally the definition of the Floer cohomology $I^*(Y)$ of Y can be seen as a Morse theoretic construction for an infinite dimensional manifold which has the homotopy type of a classyfying space of an infinite dimensional group [F].

Let $\mathcal{G}$ be the (infinite dimensional) group $Map(Y, SU_2)$, then taking degree gives a map $\deg : \mathcal{G} \longrightarrow \pi_0(\mathcal{G}) = \mathbf{Z}$. The subset $\mathcal{G}_0$ of degree 0 elements of $\mathcal{G}$ is the connected component containing the identity. The constant maps to the centre $\{\pm 1\}$ of SU_2 is contained in $\mathcal{G}_0$. The exact sequence

$$1 \longrightarrow \mathcal{G}_0/\pm 1 \longrightarrow \mathcal{G}/\pm 1 \xrightarrow{\deg} \mathbf{Z} \longrightarrow 0$$

induces maps between classifying spaces up to homotopy; there is a $\mathbf{Z}$-covering $B(\mathcal{G}_0/\pm 1) \longrightarrow B(\mathcal{G}/\pm 1)$ which is identified with the pull back of $\mathbf{R} \longrightarrow S^1$ by a map $B\deg : B(\mathcal{G}/\pm 1) \longrightarrow B\mathbf{Z} = S^1$.

A differential geometric realization of this situation is given by the space $\mathcal{A}$ of connections on $Y \times SU_2$ and the Chern-Simons invariant $cs : \mathcal{A} \longrightarrow \mathbf{R}$. Identify $\mathcal{G}$ with the gauge group of $Y \times SU_2$, and $\mathcal{G}$ naturally actson $\mathcal{A}$ so that $cs(g(A)) = cs(A) + deg(g)$. Here the Chern-Simons invariant cs is characterized as folows. The tangent space of $\mathcal{A}$ at each point is identified with the space $\Omega^1(Y, su_2)$ of 1-forms with valued in su_2. Using the paring $(u, v) \in \Omega^1(Y, su_2) \times \Omega^2(Y, su_2) \longrightarrow \int -tr(u \wedge v)$, the curvature $F : \mathcal{A} \longrightarrow \Omega^2(Y, su_2)$ can be regarded as a 2-form on $\mathcal{A}$. One can check that this is closed. The Chern-Simons invariant is the unique map $cs : \mathcal{A} \longrightarrow \mathbf{R}$ which satisfies $d(cs) = F$ and $cs(\theta_0) = 0$ for the trivial connection θ_0 on $Y \times SU_2$. Then cs descends to a map $\overline{cs} : \mathcal{B} = \mathcal{A}/\mathcal{G} \longrightarrow S^1$. Let $\mathcal{A}^*$ be the space of irreducible connections, connections whose stabili zer is ± 1. Then we have a $\mathbf{Z}$-covering $\mathcal{B}_0^* = \mathcal{A}^*/\mathcal{G}_0 \longrightarrow \mathcal{B}^* = \mathcal{A}^*/\mathcal{G}$ which is the pull back of $\mathbf{R} \longrightarrow \mathbf{Z}$ by $\overline{cs} : \mathcal{B}^* \longrightarrow S^1$. Since $\mathcal{A}$ is contractible and $\mathcal{A}^*$ is the complement of a subspace of codimension infinity, this is also weakly homotopy equivalent to a point, and the quotient spaces $\mathcal{B}^*$ and $\mathcal{B}_0^*$ are classifying spaces. Hence the Chern-Simons invariant gives the required differential geometric realization.

Formally the Floer cohomology is defined by the Morse theory for $cs : \mathcal{B}_0^* \longrightarrow \mathbf{R}$. In this article we only consider a formal aspect of the definition of the Floer cohomology and do not go into any details of analysis. Critical points of cs are the classes of flat connections. Fix a Riemannian metric on Y, which inducesa metric on $\mathcal{B}_0^*$. We assume the transversality condition as in section 1. (Since we consider only the formal aspect, we do not explain here a precise formulation of the transversality.) For two classes of irreducible flat connections $[a], [b] \in \mathcal{B}_0^*$, the space $\mathcal{M}([a],[b])$ is finite dimensional, if not empty, and the *relative Morse index* $\lambda([a],[b])$ is defined to be dim $\mathcal{M}([a],[b])$. Even if this space is empty, by using spectral flow, the relative Morse index is well-defined. Though θ_0 is not irreducible, the space $\mathcal{M}([a],[\theta_0])$ and hence $\lambda([a],[\theta_0])$ is well-defined. Then $\lambda([a]) = -\lambda([a],[\theta_0])$ satisfies $\lambda([a],[b]) = \lambda([a]) - \lambda([b])$. Like the relative indices, by using trivialization of a certain real determinant line bundle, one could define the space of *relative orientations* between $TU_{[a]}$ and $TU_{[b]}$, which consists of two points, and one could define $C([a])$ as the rank one free $\mathbf{Z}$-module spanned by the space of relative orientations between $TU_{[a]}$ and $TU_{[\theta_0]}$. However we shall only use the $\mathbf{Z}_2$-coefficientFloer cohomology, and we do not need $\mathbf{Z}$-modules. Here we simply define $C([a]) \otimes \mathbf{Z}_2$ as the free $\mathbf{Z}_2$-module generated by $[a]$. (See Remark at the end of section 1.) The $\mathbf{Z}_2$-coefficient Floer cohomology groups $I^*(Y, \mathbf{Z}_2)$ is defined as the cohomology of the differential complex $(C^* \otimes \mathbf{Z}_2, d)$, where $C^\lambda \otimes \mathbf{Z}_2 = \oplus_{\lambda([a])=\lambda} C([a]) \otimes \mathbf{Z}_2$ and d is defined as in section 1-1. The degree of the cohomology runs through $\mathbf{Z}$. The covering transformation group $\mathbf{Z}$ acts on the differential complex, which shifts the degree by mutiples of 8. Hence there is a isomorphism $\beta : I^*(Y, \mathbf{Z}_2) \longrightarrow I^{*+8}(Y, \mathbf{Z}_2)$, and this cohomology has period 8. (Any isometry preserving a Morse function on a finite dimensional manifold also preserves Morse indices of critical points. Therefore the periodicity reflects on the fact that the base space $\mathcal{B}_0^*$ is infinite dimensional.)

3-2 $H^2(\mathcal{B}_0^*, \mathbf{Z}_2)$ and $I^*_{(i)}(Y)$.

In this subsection we state some properties of a variant of the Floer cohomology which is constructed by using a cohomology class, as in sections 1 and 2.

We begin with calculating $H^2(\mathcal{B}_0^*, \mathbf{Z}_2)$. Note that

$$\pi_1(\mathcal{B}_0^*) = \pi_0(\mathcal{G}_0/\pm 1) = 0,$$

and we have the homotopy exact sequence of the covering $\mathcal{G}_0 \longrightarrow (\mathcal{G}_0/\pm 1)$:

$$1 \longrightarrow \pi_1(\mathcal{G}_0) \longrightarrow \pi_1(\mathcal{G}_0/\pm 1) \longrightarrow \{\pm 1\} \longrightarrow 1.$$

Here $\pi_1(\mathcal{G}_0) = [\tilde{S}Y, SU_2] = \pi_4(S^3) = \mathbf{Z}_2$ (we have used that Y is a homology 3-sphere.) and it can be shown that the exact sequence splits. Hence $H_2(\mathcal{B}_0^*, \mathbf{Z}) = \pi_2(\mathcal{B}_0^*) = \pi_1(\mathcal{G}_0/\pm 1) = \mathbf{Z}_2 \oplus \mathbf{Z}_2$. This implies $H^2(\mathcal{B}_0^*, \mathbf{Z}_2) = \mathbf{Z}_2 \oplus \mathbf{Z}_2$.

Let $(SU_2)_p$ is the fibre of the trivial bundle $Y \times SU_2$ over $p \in Y$ on which $\mathcal{G}$ naturally acts. Then The second Stiefel-Whitney class w of the SO_3-bundle $\mathcal{A}^* \times_{\mathcal{G}_0} (SU_2)_p$ over $\mathcal{B}_0^*$ is a non-trivial element. The covering transformation group $\mathbf{Z}$ preserves w since w is the pull back of $w_2(\mathcal{A}^* \times_{\mathcal{G}} (SU_2)_p) \in H^2(\mathcal{B}^*, \mathbf{Z}_2)$. Write the elements of $H^2(\mathcal{B}_0^*, \mathbf{Z}_2)$ as $\{0, w, u_{(0)}, u_{(1)}\}$. Then it can be shown that the generator of the covering transformation group permutes $u_{(0)}$ and $u_{(1)}$. To distinguish u_0 and u_1, we here define u_0 as follows. It can be shown that the image of $H^2(\mathcal{B}_0^* \cup \{\theta_0\}, \mathbf{Z}_2) \longrightarrow H^2(\mathcal{B}_0^*, \mathbf{Z}_2)$ is isomorphic to $\mathbf{Z}_2$. Let $u_{(0)}$ be the non-trivial element of the image. By using any of three non trivial elements w, $u_{(0)}$ and $u_{(1)}$, it is possible to construct cohomology groups of the form $H^*((C^* \oplus C^{*-1}) \otimes \mathbf{Z}_2, \tilde{d})$. Let $I^*_{(i)}(Y)$ $(* \in \mathbf{Z})$ be the cohomology groups defined by using $u_{(i)}$. Then we have:

Properties of $I^*_{(i)}(Y)$

(1) As invariants of Y, we have $\mathbf{Z}_2$-vector spaces $I^*_{(i)}$ $(* \in \mathbf{Z})$ and also a homomorphism

$$u_{(i)} : I^*(Y, \mathbf{Z}_2) \longrightarrow I^{*+2}(Y, \mathbf{Z}_2)$$

for $i \in \mathbf{Z}_2$

(2) There is a natural homomorphism

$$I^*_{(i)}(Y) \longrightarrow I^{*+8}_{(i+1)}(Y)$$

for $i \in \mathbf{Z}_2$. Hence $I^*_{(i)}(Y)$ has period 16. We also have the relation

$$\beta u_{(i)} \beta^{-1} = u_{(i+1)} : I^*(Y, \mathbf{Z}_2) \longrightarrow I^{*+2}(Y, \mathbf{Z}_2).$$

(3) There is a long exact sequence:

$$\longrightarrow I^{*-2}(Y, \mathbf{Z}_2) \xrightarrow{u_{(i)}} I^*(Y, \mathbf{Z}_2) \longrightarrow I^{*-1}_{(i)}(Y) \longrightarrow$$

for $i \in \mathbf{Z}_2$.

(4) There is a natural homomorphism $D : I^{-2}_{(0)}(Y) \longrightarrow \mathbf{Z}_2$

(5)

$$u_{(0)} u_{(1)} = u_{(1)} u_{(0)} = 0 : I^*(Y, \mathbf{Z}_2) \longrightarrow I^{*+4}(Y, \mathbf{Z}_2).$$

Properties (1) and (3) are analogies of some properties of the construction in section 1 and 2. Property (2) comes from the covering transformations which permute $u_{(0)}$ and $u_{(1)}$. Property (4) is related to the trivial connection $[\theta_0]$, which is a singular point of $\mathcal{B}_0$. Since in previous sections we have only dealed with smooth manifolds, there is no analogous situation in sections 1 or 2, although it would be possible to understand it by an analogy for finite dimensional case. Property (5) is not a very formal one. This is related to a particular nature of the classes $u_{(0)}$ and $u_{(1)}$, which we do not explain here.

Remark. The construction of R. Fintushel and R. Stern in [FS] seems to be related to our construction of $I^*_{(i)}(Y)$.

3-3 An application of $I^*_{(i)}(Y)$.

S. K. Donaldson proved:

DONALDSON'S THEOREM B. *Suppose X is a closed oriented simply connected spin 4-manifold with $b^+ = 1$. Then for any $\alpha_1, \alpha_2, \alpha_3$ and $\alpha_4 \in H_2(X, \mathbf{Z})$*

$$Q(\alpha_1, \alpha_2) Q(\alpha_3, \alpha_4) + Q(\alpha_1, \alpha_3) Q(\alpha_2, \alpha_4) + Q(\alpha_1, \alpha_4) Q(\alpha_2, \alpha_3) \equiv 0 \bmod 2$$

The above equality and an elementary algebraic consideration imply that the rank of Q is 2 and

$$Q \cong \begin{pmatrix} 0 & 1 \\ 1 & 0 \end{pmatrix}.$$

To prove this theorem, Donaldson used a $\mathbf{Z}_2$-cohomology class, which is closely related to the class $u_{(i)}$.

We can use $I^*_{(i)}(Y)$ to extend this theorem to manifolds with boundary.

THEOREM 3.1 [FFO]. *Suppose an oriented homology 3-sphere Y is the boundary of a simply connected oriented spin 4-manifold X. If $b^+ = 1$, then for $\alpha_1, \cdots, \alpha_4 \in H_2(X, \mathbf{Z})$, there is $q(X, \alpha_1, \cdots, \alpha_4) \in I^{-2}_{(0)}(Y)$ such that*

$$Q(\alpha_1, \alpha_2)Q(\alpha_3, \alpha_4) + Q(\alpha_1, \alpha_3)Q(\alpha_2, \alpha_4) + Q(\alpha_1, \alpha_4)Q(\alpha_2, \alpha_3) \equiv Dq(X, \alpha_1, \cdots, \alpha_4) \bmod 2.$$

The exact sequence of property (3) implies:

COROLLARY 3.2. *If $b^+ = 1$ and $rank(Q) \neq 2$, then $I^*(Y, \mathbf{Z}_2) \neq 0$ for $i = -2$ or -1.*

Remark For a spin manifold with $b^+ = 2$, a similar result can be obtained by using another variant of the Floer cohomology [FFO].

REFERENCES

[D] S. K. Donaldson, *Connections, cohomology, and the intersection forms of 4-manifolds*, J. Diff. Geom. **24** (1986), 275-341.

[F] A. Floer, *An instanton invariant for 3-manifolds*, Math. Phys. **118** (1988), 215-240.

[Fk] K. Fukaya, *Floer homology for oriented 3-manifolds*, preprint.

[FFO] K. Fukaya, M. Furuta, H. Ohta, in preparation.

[FS] R. Fintushel and R. Stern, *2-Torsion instanton invariants*, preprint.

7

Yang-Mills Connections of Homogeneous Bundles II

NORIHITO KOISO
COLLEGE OF GENERAL EDUCATION, OSAKA UNIVERSITY
TOYONAKA, OSAKA, 560 JAPAN

0. Introduction

Let (M, g) be a riemannian manifold, K a compact Lie group and P a principal K-bundle over M. The action integral for connections ∇ in P defined by $\mathcal{F}_{\mathrm{YM}}(\nabla) = \int |R^\nabla|^2 v_g$ is called Yang-Mills functional and a solution of Euler-Lagrange equation for Yang-Mills functional is called Yang-Mills connection.

To obtain Yang-Mills connections, it is natural to solve the corresponding "heat equation" to Yang-Mills functional and pass to limit. This method is successful in algebraic manifold category ([**1**]).

In this paper, we attack this problem in more concrete case. We assume the base manifold M is homogeneous and consider only invariant connections. Then the heat equation reduces to an ODE. When the base manifold M is compact, this method is successful by a previous paper of the author ([**3**]). Main theorems of this paper are generalizations of the previous result to non-compact cases.

We denote homogeneous objects by putting upper suffix G. We use the value $|R^\nabla|^2$ at the origin as a Yang-Mills functional, because the manifold M may be non-compact. We denote by $\mathcal{C}^G$ the set of all invariant connections and define a function $\mathcal{F}^G_{\mathrm{YM}}$ on $\mathcal{C}^G$ by $\mathcal{F}^G_{\mathrm{YM}}(\nabla) = |R^\nabla|^2$. It is easy to see that invariant Yang-Mills connections are nothing but critical points of the function $\mathcal{F}^G_{\mathrm{YM}}$.

THEOREM A. *Let ∇_t be the solution of the ODE:*

$$\frac{d}{dt}\nabla_t = -\operatorname{grad}\mathcal{F}^G_{\mathrm{YM}}(\nabla_t)$$

with initial data ∇_0 in $\mathcal{C}^G$. Then ∇_t converges to a G-invariant Yang-Mills connection.

THEOREM B (Mountain pass lemma). *Let ∇_1 and ∇_2 be G-invariant Yang-Mills connections in P which are local minima of the function $\mathcal{F}^G_{\mathrm{YM}}$. Assume that there is no curve joining ∇_1 and ∇_2 in $\mathcal{C}^G$ consisting of Yang-Mills connections. Then there exists a G-invariant Yang-Mills connection in P which is not local minimum of the function $\mathcal{F}^G_{\mathrm{YM}}$.*

1. Basic facts

Let M be a homogeneous riemannian manifold G/H, K a compact Lie group, $\rho\colon H \to K$ a Lie group homomorphism and P the principal K-bundle $G\times_\rho K$ over M. We denote by $\mathcal{C}^G$ the set of all G-invariant connections in the principal bundle $P \to M$, and assume $\mathcal{C}^G$ is not empty. We fix a bi-invariant inner product on the Lie algebra $\mathfrak{k}$ of K and define a function $\mathcal{F}^G_{\mathrm{YM}}$ on $\mathcal{C}^G$ by $\mathcal{F}^G_{\mathrm{YM}}(\nabla) = |R^\nabla|^2$.

To analyze the function $\mathcal{F}^G_{\mathrm{YM}}$ on the space $\mathcal{C}^G$, we replace the isotropy group H by a compact Lie group by the following way. Let G_1 be the quotient Lie group of G by the kernel of the representation $G \to \mathrm{Auto}(P)$, and G' the closure of G_1 in $\mathrm{Auto}(P)$. Then G' acts transitively on M as isometries.

Let H' be its isotropy group. The homomorphism $\rho\colon H \to K$ induces a homomorphism $\rho'\colon H' \to K$. We get new objects $\{G' \times_{\rho'} K, \mathcal{C}^{G'}, \mathcal{F}^{G'}_{\mathrm{YM}}\}$, but they are canonically isomorphic to the original objects $\{G \times_\rho K, \mathcal{C}^G, \mathcal{F}^G_{\mathrm{YM}}\}$.

Moreover, the isotropy group H' is a compact Lie group. In fact, if we put a G'-invariant connection ∇ on $P \to M$ and define a riemannian metric on P by the usual way, then the group G' becomes a closed subgroup of the isometry group of P. An element γ of G' belongs to H' if and only if it preserves the fiber P_o at the origin $o \in M$. Since the fiber P_o is compact, the group H' is compact.

From now on, we replace the original objects by the new objects. In particular, the isotropy group H is assumed to be compact.

We denote by $\mathfrak{g}$, $\mathfrak{h}$ the Lie algebra of G, H respectively, and decompose $\mathfrak{g}$ into $\mathfrak{h} + \mathfrak{m}$ as H-module. The vector space $\mathfrak{m}$ is identified with the tangent space of the riemannian manifold M at the origin o. We denote by $\mathfrak{m}_0$ and $\mathfrak{m}_1$ the trivial factor and the sum of irreducible factors of the $\mathfrak{h}$-module $\mathfrak{m}$ respectively. We fix an H-invariant inner product on $\mathfrak{g}$, so that $\mathfrak{h}$ is orthogonal to $\mathfrak{m}$. The bi-invariant inner product on $\mathfrak{k}$ and the inner product on $\mathfrak{g}$ will be denoted by $\langle *, * \rangle$.

We regard the space $\mathfrak{k}$ as an H-module via the homomorphism $\rho : H \to K$. For elements $A, B \in \mathrm{Hom}_H(\mathfrak{m}, \mathfrak{k})$, we define an element $[A \wedge B] \in \mathrm{Hom}_H(\wedge^2 \mathfrak{m}, \mathfrak{k})$ by

$$[A \wedge B](v, w) = \frac{1}{2}([A(v), B(w)] + [B(v), A(w)]).$$

LEMMA 1.1 ([**2**, Chap II Theorem 11.7]). *There is a one-to-one correspondence between the set $\mathcal{C}^G$ of G-invariant connections in P and the space $\mathrm{Hom}_H(\mathfrak{m}, \mathfrak{k})$ of H-module homomorphisms from $\mathfrak{m}$ to $\mathfrak{k}$. The curvature tensor R^A of the G-invariant connection corresponding to $A \in \mathrm{Hom}_H(\mathfrak{m}, \mathfrak{k})$ is given by*

$$R^A(v, w) = [A \wedge A](v, w) - A([v, w]_{\mathfrak{m}}) - \rho([v, w]_{\mathfrak{h}}),$$

for $v, w \in \mathfrak{m}$, where lower suffixes $\mathfrak{m}$, $\mathfrak{h}$ denote the factors with respect to the decomposition: $\mathfrak{g} = \mathfrak{h} + \mathfrak{m}$.

We take orthonormal basis $\{v_i\}$ of $\mathfrak{m}$ and basis $\{v_s\}$ of $\mathfrak{h}$, and put $A_i = A(v_i)$ for $A \in \mathrm{Hom}_H(\mathfrak{m}, \mathfrak{k})$ and $\rho_s = \rho(v_s)$. Structure constants of the Lie algebra $\mathfrak{g}$ is denoted by $C_*{}^*{}_*$. In particular, $[v_i, v_j] = \sum C_i{}^k{}_j v_k + \sum C_i{}^s{}_j v_s$, where the summation is taken for indexes which appear twice. We use this convention from now on.

LEMMA 1.2 ([**3**, (2.2)]). *The gradient vector* $\mathrm{grad}\,\mathcal{F}^G_{\mathrm{YM}}$ *at* $A \in \mathrm{Hom}_H(\mathfrak{m}, \mathfrak{k})$ *is given by*

$$(\mathrm{grad}\mathcal{F}^G_{\mathrm{YM}})(v_i) = -\sum [A_j, [A_j, A_i]] + \sum C_j{}^k{}_i [A_j, A_k] + \sum C_j{}^s{}_i [A_j, \rho_s] \\ - \frac{1}{2} \sum C_j{}^i{}_k [A_j, A_k] + \frac{1}{2} \sum C_j{}^i{}_k C_j{}^l{}_k A_l + \frac{1}{2} \sum C_j{}^i{}_k C_j{}^s{}_k \rho_s.$$

Remark that we may assume the Lie algebra $\mathfrak{k}$ is abelian or semi-simple. In fact, when the algebra $\mathfrak{k}$ decomposes into the abelian factor and the semi-simple factor, the space $\mathrm{Hom}_H(\mathfrak{m}, \mathfrak{k})$, the function $\mathcal{F}^G_{\mathrm{YM}}$ and the vector field $\mathrm{grad}\,\mathcal{F}^G_{\mathrm{YM}}$ decomposes correspondingly, and the proof of Theorem A and B reduces to each case.

2. Proof of Theorems

LEMMA 2.1 (c.f. [**3**, Lemma 1.4]). *There exist positive constants* c_1, c_2 *and* c_3 *such that for any* $A \in \mathrm{Hom}_H(\mathfrak{m}, \mathfrak{k})$ *it holds that*

$$|R^A| \geq c_1 \big|A|\mathfrak{m}_1\big|^2 - c_2 \big|A|\mathfrak{m}_0\big| - c_3.$$

Proof. If $[A \wedge A] = 0$, then

$$0 = \langle [A(\mathfrak{m}), A(\mathfrak{m})], \rho(\mathfrak{h}) \rangle = \langle A(\mathfrak{m}), [\rho(\mathfrak{h}), A(\mathfrak{m})] \rangle \\ = \langle A(\mathfrak{m}), A([\mathfrak{h}, \mathfrak{m}]) \rangle = \langle A(\mathfrak{m}), A(\mathfrak{m}_1) \rangle.$$

Therefore if we set $c_1 = \inf\{|[A \wedge A]|; A \in \mathrm{Hom}_H(\mathfrak{m}_1, \mathfrak{k}), |A| = 1\}$, then $c_1 > 0$ and the desired inequality follows from Lemma 1.1. Q.E.D.

REMARK. When the trivial factor $\mathfrak{m}_0$ vanishes, this estimate immediately leads the boundedness of solutions of the gradient flow and the compactness of the space of G-invariant Yang-Mills connections.

It is shown as follows. We denote by $A_{\mathfrak{m}}$ a $\mathfrak{k}$-valued 2-form on $\mathfrak{m}$ defined by $A_{\mathfrak{m}}(v, w) = A([v, w]_{\mathfrak{m}})$. Assume that the space $\mathfrak{m}_0$ vanishes and that the value $|A|$ is sufficiently large. Then the quantity

$$\frac{1}{2} \frac{d}{dt} |_{t=0} |R^{(A+tA)}|^2 = \langle R^A, 2[A \wedge A] - A_{\mathfrak{m}} \rangle$$

is strictly positive by the above estimate. It implies that the norm $|A|$ is strictly decreasing along the gradient flow outside certain compact set.

Let $A(t)$ be a solution of the equation $dA(t)/dt = -\mathrm{grad}\,\mathcal{F}^G_{\mathrm{YM}}(A(t))$. To show the boundedness of $A(t)$, it suffices to prove that $A(t)|\mathfrak{m}_0$ is bounded, because boundedness of $A|\mathfrak{m}_0$ implies boundedness of $A|\mathfrak{m}_1$ by the previous lemma.

To prove it, we have to choose a good orthonormal basis $\{v_i\}$ of $\mathfrak{m}$. Let ϕ be a symmetric bilinear form on $\mathfrak{m}$ defined by $\phi(v_i, v_j) = \sum C_k{}^i{}_l C_k{}^j{}_l$. Since ϕ is an H-invariant form, we can choose orthonormal basis $\{v_i\}$ of $\mathfrak{m}_0$ and orthonormal basis $\{v_j\}$ of $\mathfrak{m}_1$ respectively so that ϕ becomes a diagonal matrix.

We fix this basis, choose any vector $v_i \in \mathfrak{m}_0$ and denote it by v_0. The vector in Lemma 1.2 is decomposed into 3 parts as follows.

$$\Phi^1 = -\sum [A_j, [A_j, A_0]] + \sum C_j{}^k{}_0 [A_j, A_k] - \frac{1}{2}\sum C_j{}^0{}_k [A_j, A_k],$$
$$\Phi^2 = \frac{1}{2}\sum C_j{}^0{}_k C_j{}^0{}_k A_0 + \frac{1}{2}\sum C_j{}^0{}_k C_j{}^s{}_k \rho_s,$$
$$\Phi^3 = \sum C_j{}^s{}_0 [A_j, \rho_s].$$

LEMMA 2.2. $\Phi^3 = 0$.

Proof. First, we observe that

$$\sum C_j{}^s{}_0 [A_j, \rho_s] = \sum [A_j, \rho([v_j, v_0]_{\mathfrak{h}})]$$
$$= \sum [\rho([v_0, v_j]_{\mathfrak{h}}), A(v_j)] = A(\sum [[v_0, v_j]_{\mathfrak{h}}, v_j]).$$

Here, the correspondence: $v_i \mapsto \sum [[v_i, v_j]_{\mathfrak{h}}, v_j]$ generates an $\mathfrak{h}$-endmorphsm of $\mathfrak{m}$. Therefore, when v_i belongs to $\mathfrak{m}_0$, we may take the sum as v_j runs through the basis of $\mathfrak{m}_0$. Thus $\sum [[v_0, v_j]_{\mathfrak{h}}, v_j]$ vanishes. Q.E.D.

LEMMA 2.3. *If $\mathfrak{k}$ is abelian then $A_0(t)$ is bounded.*

Proof. When the algebra $\mathfrak{k}$ is abelian, Φ^1 also vanishes. Therefore,

$$\frac{d}{dt}|A_0(t)|^2 = -\sum C_j{}^0{}_k C_j{}^0{}_k |A_0(t)|^2 - \sum C_j{}^0{}_k C_j{}^s{}_k \langle \rho_s, A_0(t)\rangle.$$

If all $C_j{}^0{}_k$ vanish, then $dA_0(t)/dt = 0$. If there is a non-zero $C_j{}^0{}_k$, then $d|A_0(t)|^2/dt$ is negative for sufficiently large $|A_0(t)|$. Q.E.D.

In the following, we assume that K is semi-simple and introduce the following usual notations. Let $\mathfrak{t}$ be a Cartan subalgebra of the Lie algebra $\mathfrak{k}$ and Δ its root system as a subset of $\mathfrak{t}$. We characterize root vectors $X_\alpha \in \mathfrak{t}$ corresponding to $\alpha \in \Delta$ by (1) $[u, X_\alpha] = \langle u, \alpha\rangle X_{-\alpha}$ for all $u \in \mathfrak{t}$ and (2) $[X_\alpha, X_{-\alpha}] = \alpha$. We denote by $\{\alpha_p\}$ the fundamental root system and by $\{\omega_p\}$ the fundamental weight system. Let 2δ be the sum of all positive roots. We put $2\delta = \sum n_p \alpha_p$ and set $\xi_p = 1/n_p \cdot \omega_p$. For $w \in \mathfrak{k}$, we define

$$L(w) = \max\{\langle \mathrm{Ad}_\gamma w, \xi_p\rangle; 1 \le p \le \dim \mathfrak{t}, \gamma \in K\}.$$

LEMMA 2.4 ([**3**, Lemma 2.1, 2.2]). *For $w \in \mathfrak{k}$, the value $L(w)$ is realized by $\gamma \in K$ such that $\mathrm{Ad}_\gamma w$ belongs to the positive Weyl chamber $\overline{W}$. In particular L is a norm of $\mathfrak{k}$. Moreover, there is a positive constant ϵ such that if w belongs to $\overline{W}$ and $L(w)$ is realized by ξ_p then it holds that $\langle w, \alpha_p\rangle \ge \epsilon L(w)$.*

Since $A_0(t)$ is real analytic, $L(A_0(t))$ is continuous and, by the previous lemma, is piecewise represented as $L(A_0(t)) = \langle \mathrm{Ad}_{\gamma(t)} A_0(t), \xi_1\rangle$, $Ad_{\gamma(t)} A_0(t) \in \overline{W}$, where $\gamma(t)$ is a real analytic curve of K and ξ_1 is taken by renumbering of suffix. We may assume that $\gamma(t)$ is the identity at a time t_0 by changing the Cartan subalgebra of $\mathfrak{k}$. Therefore, to show Theorem A, it is sufficient to prove that $\langle \Phi^2, \xi_1\rangle$ and $\langle \Phi^1, \xi_1\rangle$ are non-negative under the assumption.

LEMMA 2.5. *If $L(A_0)$ is sufficiently large, then $\langle \Phi^2, \xi_1 \rangle$ is non-negative.*

Proof. If all $C_j{}^0{}_k$ vanish, then $\Phi^2 = 0$. If there is a non-zero $C_j{}^0{}_k$, then

$$\langle \Phi^2, \xi_1 \rangle = \frac{1}{2} \sum C_j{}^0{}_k C_j{}^0{}_k L(A_0) + \frac{1}{2} \sum C_j{}^0{}_k C_j{}^s{}_k \langle \rho_s, \xi_1 \rangle$$

is positive for sufficiently large $L(A_0)$. Q.E.D.

LEMMA 2.6. *If $L(A_0)$ is sufficiently large, then $\langle \Phi^1, \xi_1 \rangle$ is non-negative.*

Proof. Put $A_j = u_j + \sum x_j^\alpha X_\alpha$, where $u_j \in \mathfrak{t}$ and $\alpha \in \Delta$. Express Φ^1 as

$$-\sum [A_j, [A_j, A_0]] + \sum c_{jk}[A_j, A_k]$$

combining the last two terms. Then,

$$\begin{aligned}
\langle \Phi^1, \xi_1 \rangle &= \sum \langle [A_0, A_j], [\xi_1, A_j] \rangle + \sum c_{jk} \langle [\xi_1, A_j], A_k \rangle \\
&= \sum x_j^\alpha x_j^\alpha |X_\alpha|^2 \langle \xi_1, \alpha \rangle \langle \alpha, A_0 \rangle + \sum c_{jk} x_j^\alpha x_k^{-\alpha} |X_\alpha|^2 \langle \xi_1, \alpha \rangle \\
&= \sum{}^+ x_j^\alpha x_j^\alpha |X_\alpha|^2 \langle \xi_1, \alpha \rangle \langle \alpha, A_0 \rangle + \sum{}^+ x_k^{-\alpha} x_k^{-\alpha} |X_\alpha|^2 \langle \xi_1, \alpha \rangle \langle \alpha, A_0 \rangle \\
&\quad + \sum{}^+ c_{jk} x_j^\alpha x_k^{-\alpha} |X_\alpha|^2 \langle \xi_1, \alpha \rangle - \sum{}^+ c_{kj} x_k^{-\alpha} x_j^\alpha |X_\alpha|^2 \langle \xi_1, \alpha \rangle,
\end{aligned}$$

where the summation $\sum^+$ is taken for positive roots α. More strictly, this summation is taken only for positive roots $\alpha \in \Delta$ such that $\langle \alpha, \xi_1 \rangle \neq 0$. If we represent such α as $\sum m_p \alpha_p$, then $m_p \geq 1$ and all $m_p \geq 0$. Therefore, $\langle \alpha, A_0 \rangle \geq \langle \alpha_1, A_0 \rangle \geq \epsilon L(A_0)$ by Lemma 2.4. We regard the last expression as a quadratic form of x_j^α and $x_k^{-\alpha}$. Then the above observation implies that the coefficients of $(x_j^\alpha)^2$ and $(x_k^{-\alpha})^2$ are sufficiently larger than that of $x_j^\alpha x_k^{-\alpha}$. Q.E.D.

Proof of Theorem A. Combining these estimates, we see that the curve $A(t)$ is bounded in $\mathrm{Hom}_H(\mathfrak{m}, \mathfrak{k})$. Since any bounded solution of real analytic gradient flow converges ([**4**]), the proof of Theorem A is finished. Q.E.D.

Proof of Theorem B. We decompose the Lie algebra $\mathfrak{k}$ into the sum of the center $\mathfrak{z}$ and the semi-simple part $\mathfrak{k}'$. We take a sufficiently large constant c and define a convex subset $\overline{S}$ of $\mathrm{Hom}_H(\mathfrak{m}, \mathfrak{k})$ by

$$\{A \in \mathrm{Hom}_H(\mathfrak{m}, \mathfrak{k}); L((A(v_i))_{\mathfrak{k}'}) \leq c \text{ and } |(A(v_i))_{\mathfrak{z}}| \leq c \text{ for all } v_i \in \mathfrak{m}_0\}.$$

Then Lemma 2.1 implies that the function $\mathcal{F}_{\mathrm{YM}}^G$ is proper on the set $\overline{S}$. Moreover, the proof of Lemma 2.3, Lemma 2.5 and 2.6 implies that the gradient vector is not outward on the boundary of the set $\overline{S}$. Thus Struwe's Mountain pass lemma [**5**] leads our result. Q.E.D.

References

[1] S.K. Donaldson: *Infinite determinants, stable bundles and curvature*, Duke Math. J. **54** (1987), 231–247.

[2] S. Kobayashi and K. Nomizu: Foundations of Differential Geometry, Vol. I, Interscience publishers, New York, 1963.

[3] N. Koiso: *Yang-Mills connections of homogeneous bundles*, Osaka J. Math. **27** (1990), 163–174.

[4] S. Łojasiewicz, Jr.: *Sur les trajectoires du gradient d'une fonction analytique.*, Geometry seminars, 1982–1983 (1984), 115–117, Univ. Stud. Bologna.

[5] M. Struwe: *Plateau's problem and the calculus of variations*, Vorlesungsreiche SFB 72 No. 32 (1986), Bonn University.

8

Non-Trivial Harmonic Spinors on Certain Algebraic Surfaces

D. Kotschick
Queens' College, Cambridge, England
and
Mathematisches Institut, Universität Basel,
Basel, Switzerland

1. Introduction

The application of Yang-Mills theory to 4-dimensional differential topology has led to a renewed interest in spaces of (twisted) harmonic spinors on 4-manifolds and on algebraic surfaces [Do2]. These spaces depend on a choice of Riemannian metric, but, for a Kähler metric turn out to be invariants of the complex structure [Hi]. Nevertheless, specially adapted metrics can be used to do explicit computations. It may thus be appropriate to report some new interesting examples of harmonic spinors on algebraic surfaces in a volume devoted to Yang-Mills connections and Einstein metrics.

The aim of this paper is to show that certain simply-connected spin algebraic surfaces carry spaces of harmonic spinors whose dimension is larger than the index formula for the Dirac operator predicts. To the author's knowledge, these are the first such examples. We will say that such a surface has non-trivial harmonic spinors.

Hitchin [Hi] showed that simply-connected Kähler surfaces which either are not of general type, or are complete intersections or cyclic covers of $\mathbb{C}P^2$ branched in a smooth curve do not carry non-trivial harmonic spinors. In all these cases the index of the Dirac operator is positive and there are no negative harmonic spinors. When the index is negative, one might expect, *mutatis mutandis*, that there are no positive harmonic spinors. We give counterexamples to this conjecture.

The examples we present are the surfaces of positive signature constructed by Moishezon and Teicher [MT1], [MT2], cf. also [Mi]. For these the index formula forces the existence of negative harmonic spinors, but the surfaces have positive harmonic spinors as well. There is also a zero-signature example with non-trivial harmonic spinors. Thus, our result is:

Theorem 1. *The simply-connected Moishezon-Teicher surfaces are spin. All those with non-negative signature carry non-trivial harmonic spinors.*

In fact the dimension of the space of non-trivial harmonic spinors can be arbitrarily large.

By taking products of these surfaces and using the Künneth formula for harmonic spinors [Hi], one finds examples of simply-connected Kähler manifolds with arbitrarily large spaces of non-trivial harmonic spinors in all even complex dimensions.

One does not immediately obtain examples in odd complex dimensions because taking products with $\mathbb{C}P^1$ kills the harmonic spinors, as it should, because the product has positive scalar curvature. However, we give examples in odd complex dimensions larger than one at the end of this note. Thus, we obtain:

Corollary 1. *In every complex dimension > 1 there exist simply-connected Kähler manifolds with arbitrarily large spaces of non-trivial harmonic spinors.*

2. Harmonic spinors

Let X be a closed oriented Riemannian spin manifold. The Dirac operator is an elliptic differential operator acting on sections of the spin bundle $V \to X$. Its (finite-dimensional) kernel is the space H of harmonic spinors. If the dimension of X is even, then the Dirac operator interchanges the bundles of positive and of negative spinors in the orthogonal decomposition $V = V_+ \oplus V_-$ and one considers the (half-) Dirac operator

$$P : \Gamma(V_+) \longrightarrow \Gamma(V_-).$$

One has $ker(P) = H^+$, $coker(P) = H^-$; where $H = H^+ \oplus H^-$ is the splitting induced by the splitting of V. The index of P is given by the (generalised) $\hat{A}$-genus. In dimension 4 this reduces to

$$index(P) = -\frac{1}{8} signature(X). \tag{1}$$

One always has $dim H \geq |index(P)|$, so that the index formula gives a lower bound for the number of independent harmonic spinors in terms of the topology of X.

For the case when the Riemannian metric is Kähler, Hodge theory gives the following interpretation of harmonic spinors [Hi]:

Theorem 2. *On a Kähler spin manifold X there are canonical isomorphisms*

$$H^+ = H^{even}(X, \mathcal{O}(\frac{1}{2}K)),$$

$$H^- = H^{odd}(X, \mathcal{O}(\frac{1}{2}K)).$$

Here $\frac{1}{2}K$ denotes a half-canonical divisor, the existence of which is equivalent to X being spin. This result shows that, while the dimension of H may (and does) vary in a general family of Riemannian metrics, it is constant in any family of Kähler metrics compatible with a fixed complex structure. The index formula for P specializes to the Riemann-Roch theorem for $\frac{1}{2}K$. In complex dimension 2 it is

$$\chi(X, \mathcal{O}(\frac{1}{2}K)) = \frac{1}{24}(2c_2(X) - c_1^2(X)), \tag{2}$$

which checks with (1), of course. Note that $H^i(X, \mathcal{O}(\frac{1}{2}K))$ and $H^{2-i}(X, \mathcal{O}(\frac{1}{2}K))$ are Serre dual, so the dimensions of H^0 and of H^2 agree.

Thus one has a topological lower bound on the amount of cohomology for $\frac{1}{2}K$, and it is natural to ask when there is more. Hitchin studied the case of complex dimension 1, where there are examples of excess cohomology for curves of genus $g > 2$. One can construct examples in any dimension by taking products of curves, but they will not be simply connected, of course. We will give examples of simply-connected surfaces with excess cohomology of $\frac{1}{2}K$ in the next section.

3. Non-trivial examples

Denote by X the quadric surface $\mathbb{C}P^1 \times \mathbb{C}P^1$, and by l_1, l_2 the two factors. Consider the projective embedding $X_{a,b}$ of X defined by the linear system $|al_1 + bl_2|$, where a and b are positive integers. A generic projection $f : X_{a,b} \to \mathbb{C}P^2$ is ramified in a curve B of degree $6ab - 2a - 2b$ with only ordinary double points and (2,3)-cusps as its singularities. Now define $g : Y_{a,b} \to \mathbb{C}P^2$ to be the Galois closure of f. By the calculation in [MT2], one has

$$signature(Y_{a,b}) = \frac{1}{3}((2ab)!)(ab - 3a - 3b + 5). \tag{3}$$

The following deep result was proved in [MT2]:

Theorem 3. *If a and b are relatively prime, then $Y_{a,b}$ is simply-connected.*

We now prove Theorem 1. The branching order of g at a generic point of B is 2, so the canonical class of $Y_{a,b}$ is

$$K_{Y_{a,b}} = g^*(K_{\mathbb{C}P^2} + \frac{1}{2}B) = (3ab - a - b - 3)g^*(H),$$

where H is the hyperplane class of the plane. If $(a, b) = 1$, then $3ab - a - b - 3$ is even and hence $Y_{a,b}$ is spin. Moreover,

$$g^* : H^0(\mathbb{C}P^2, \mathcal{O}(\frac{1}{2}(3ab - a - b - 3)H)) \longrightarrow H^0(Y_{a,b}, \mathcal{O}(\frac{1}{2}K_{Y_{a,b}}))$$

is an injection, so using Serre duality and Theorem 2 we obtain

$$dimH^+ \geq 2dimH^0(\mathbb{C}P^2, \mathcal{O}(\frac{1}{2}(3ab-a-b-3)H)) = \frac{1}{4}((3ab-a-b)^2-1) > 0.$$

This lower bound becomes arbitrarily large as a and b are made large. On the other hand, if $a \geq 6$ and $b \geq 5$, or if $a \geq 7$ and $b = 4$, the signature is non-negative by (3) and (1) does not predict any positive harmonic spinors. This completes the proof of Theorem 1.

The surface $Y_{7,4}$ is particularly interesting. It has zero signature, so by [Fr] and [Do1] it is homeomorphic but not diffeomorphic to a connected sum of quadrics. This connected sum carries metrics of positive scalar curvature, but $Y_{7,4}$ is not known to do so. Even without the presence of harmonic spinors, it would be clear that no Kähler metric on $Y_{7,4}$ can have positive scalar curvature: whenever there are non-vanishing plurigenera the total scalar curvature of any Kähler metric is negative. Note that in higher dimensions the vanishing of the generalized $\hat{A}$-invariant is sufficient for simply-connected spin manifolds to carry metrics of positive scalar curvature [St].

For all the surfaces considered here the canonical bundle is ample, so they carry Kähler-Einstein metrics of negative scalar curvature.

Finally, we prove Corollary 1. If X is spin of odd complex dimension, then Serre duality implies $\chi(X, \mathcal{O}(\frac{1}{2}K_X)) = 0$. However, if X is a hypersurface of degree $2k+1$ in $\mathbb{C}P^{2n}$, then it is spin and $\mathcal{O}(\frac{1}{2}K_X) = \mathcal{O}_{\mathbb{C}P^{2n}}(k-n)|_X$ has an arbitrarily large space of sections for large k. For $n > 1$, X is simply-connected.

References

[Do1] S.K. Donaldson: Polynomial invariants for smooth four-manifolds. *Topology* **29** (1990), 257–315.

[Do2] S.K. Donaldson: Differential Topology and Complex Variables. in Proceedings of the 29th *Mathematischc Arbcitstagung*, MPI Bonn 1990.

[Fr] M.H. Freedman: The topology of four-dimensional manifolds. *J. Differential Geometry* **17** (1982), 357–453.

[Hi] N.J. Hitchin: Harmonic spinors. *Adv. in Math.* **14** (1974), 1–55.

[Mi] Y. Miyaoka: Algebraic surfaces with positive indices. in *Classification of Algebraic and Analytic Manifolds*, Birkhäuser Verlag, Basel, Boston 1982.

[MT1] B. Moishezon and M. Teicher: Existence of simply connected algebraic surfaces of general type with positive and zero indices. *Proc. Natl. Acad. Sci. USA* **83** (1986), 6665–6666.

[MT2] B. Moishezon and M. Teicher: Simply-connected algebraic surfaces of positive index. *Invent. Math.* **89** (1987), 601–643.

[St] S. Stolz: Simply connected manifolds of positive scalar curvature. *Bulletin AMS* **23** (1990), 427–432.

9
Cohomology on Symmetric Products, Syzygies of Canonical Curves, and a Theorem of Kempf

Robert Lazarsfeld*
Department of Mathematics
U.C.L.A.
Los Angeles, CA 90024

Let C be a compact connected Riemann surface of genus $g \geq 2$, denote by C_m the mth symmetric product of C, and consider the Abel-Jacobi map

$$u : C_m \longrightarrow J(C),$$

where $J(C)$ is the Jacobian of C. The derivative of u determines a homomorphism

$$du : \Theta_{C_m} \longrightarrow u^* \Theta_{J(C)}$$

of coherent sheaves on C_m, where as usual Θ_X denotes the tangent sheaf of a complex manifold X. In the course of his celebrated investigation [K1] of the deformation theory of symmetric products, Kempf computed $H^1(C_m, \Theta_{C_m})$ and analyzed the map $H^1(du)$ induced by du. His result is that $H^1(C_m, \Theta_{C_m}) = H^1(C, \Theta_C)$, and that $H^1(du)$ is identified with the canonical homomorphism $H^1(C, \Theta_C) \longrightarrow H^1(C, \Theta_C) \otimes H^1(C, \Theta_C)$ dual to the multiplication $H^0(C, \Omega) \otimes H^0(C, \Omega) \longrightarrow H^0(C, \Omega^2)$, Ω being the canonical bundle on C. In particular, it then follows from Noether's theorem that $H^1(du)$ fails to be injective if and only if C is hyperelliptic.

The purpose of this note is to carry out the analogous computations for higher cohomology groups. Surprisingly, we find that the answer involves the syzygies of canonical curves.

To give precise statements, we start with some notation. Let $M = M_\Omega$ denote the kernel of the evaluation homomorphism $H^0(C, \Omega) \otimes \mathcal{O}_C \longrightarrow \Omega$, and write $Q = Q_\Omega = M^*$, so that Q is a vector bundle of rank $g-1$ on X. Making the identification $H^1(C, \Theta_C) = H^0(C, \Omega)^*$, one thus has an exact sequence

$$0 \longrightarrow \Theta_C \longrightarrow H^1(C, \Theta_C) \otimes \mathcal{O}_C \longrightarrow Q \longrightarrow 0$$

---

* Partially Supported by NSF Grant DMS 89-02551

of vector bundles on C, which in turn gives rise to

$$(*)_k \qquad 0 \longrightarrow \Lambda^{k-1}Q \otimes \mathcal{O}_C \longrightarrow \Lambda^k H^1(C, \mathcal{O}_C) \otimes \mathcal{O}_C \longrightarrow \Lambda^k Q \longrightarrow 0.$$

In technical terms, we may summarize the result of our computation in the following

Theorem. *If* $m \geq k$ *and* $k < g-1$, *then there is a canonical isomorphism*

$$H^k(C_m, \mathcal{O}_{C_m}) = H^1(C, \Lambda^{k-1}Q \otimes \mathcal{O}_C),$$

and the map $H^k(du) : H^k(C_m, \mathcal{O}_{C_m}) \longrightarrow H^k(C_m, u^* \mathcal{O}_{J(C)})$ *is identified with the homomorphism* $H^1(C, \Lambda^{k-1}Q \otimes \mathcal{O}_C) \longrightarrow \Lambda^k H^1(C, \mathcal{O}_C) \otimes H^1(C, \mathcal{O}_C)$ *determined by* $(*)_k$.

A more picturesque formulation of this result involves the syzygies of the canonical embedding $C \subset \mathbb{P}^{g-1}$ of C. Recall (c.f. [GL1] or [L1]) that one says that the canonical bundle Ω satisfies **property** $(\mathbf{N_p})$ if roughly speaking the first p steps in the minimal graded free resolution of the homogeneous ideal of $C \subset \mathbb{P}^{g-1}$ are as simple as possible. Referring to [GL1] or [L1] for the precise definition, suffice it to say here that Ω satisfies (N_0) iff Ω is normally generated; (N_1) holds iff Ω is normally generated and in addition the homogeneous ideal $I_{C/\mathbb{P}^{g-1}}$ is generated by quadrics; (N_2) holds iff (N_1) does, and the module of syzygies among quadratic generators $Q_i \in I_{C/\mathbb{P}^{g-1}}$ is generated by relations of the form $\Sigma L_i Q_i = 0$ where the L_i are *linear* polynomials; and so on. A well-known conjecture of Green's [G1] asserts that least value of p for which (N_p) fails for Ω is equal to the Clifford index Cliff(C) of C. (See [L2] for a recent survey of this and related conjectures.)

It is standard and elementary (consult [GL2], [L1] or [PR]) that the syzygies of Ω are governed by the exact sequences $(*)_k$. Specifically, (N_k) holds for Ω if and only if the homomorphism $H^1(C, \Lambda^k Q \otimes \mathcal{O}_C) \longrightarrow \Lambda^{k+1} H^1(C, \mathcal{O}_C) \otimes H^1(C, \mathcal{O}_C)$ determined by $(*)_{k+1}$ is injective. Hence one has the

Corollary. *If* $k \leq m$ *and* $k < g-1$, *then* $H^k(du)$ *fails to be injective if and only if property* (N_{k-1}) *fails for the canonical bundle* Ω.

So for example, $H^2(du)$ fails to be injective if and only if C is Petri-exceptional. The Corollary helps to explain some of the computations appearing in [K2] and [Muk]. We hope that it may eventually open the door to some progress on computing the syzygies of generic canonical curves. Some other, more geometric, variants of the Theorem and its Corollary appear in §3.

The reader will recognize my debt to Kempf's paper [K1]. I am also grateful to L. Ein, M. Green, G. Kempf, M. Schacher and C. Voisin for valuable discussions.

§1. A Lemma on Cohomology and Galois Coverings

The purpose of this section is to record an elementary result (Proposition 1.1) concerning invariant cohomology classes on Galois coverings in characteristic zero. The fact in question is certainly known in much greater generality (c.f. [Tohuku]), but we have been unable to find a suitable reference for the particular statement we need. Therefore, for the benefit of the reader, we give here an elementary direct argument.

Let $f : X \longrightarrow Y$ be a finite surjective mapping of complex algebraic varieties. Recall that one says that f is Galois with (finite) group G if G acts on X by automorphisms commuting with f, and if the sheaf $\mathcal{O}_Y$ of germs of functions on Y consists of the G-invariant germs of sections of $\mathcal{O}_X$. In other words, $Y = X/G$.

Proposition 1.1. Let $f : X \longrightarrow Y$ be a Galois covering of complex algebraic varieties with group G, and set $A = \Gamma(Y, \mathcal{O}_Y)$. Let L be a line bundle on Y, so that f^*L is a G-bundle on X, and consider the corresponding action of G on the A-module $H^i(X, f^*L)$. Then there is a canonical isomorphism

$$H^i(Y, L) \xrightarrow{\simeq} H^i(X, f^*L)^G,$$

where the space on the right is the submodule of invariants of the G-module $H^i(X, f^*L)$.

Proof. Since f is finite (and hence affine) we have using the projection formula a canonical identification

$$(*) \qquad H^i(X, f^*L) = H^i(Y, L \otimes f_* \mathcal{O}_X),$$

with G- module structure arising from the natural action of G on the sheaf $f_*\mathcal{O}_X$ on Y. Choose an affine open covering $\underline{U} = \{ U_\alpha \}$ of Y, and let $C = C^{\cdot}(\underline{U}, L \otimes f_*\mathcal{O}_X)$ be the corresponding Cech complex of A-modules computing the groups on the right in (*). G acts on this complex by chain homomorphisms. Since $\mathcal{O}_Y = f_*\mathcal{O}_X{}^G$, we see that $H^i(Y, L) = H^i(C^G)$, where $C^G \subset C$ denotes the subcomplex of G-invariants. Therefore the assertion of the Proposition boils down to the statement that the canonical map $H^i(C^G) \longrightarrow H^i(C)^G$ is an isomorphism. But this is the content of the following Lemma.∎

Lemma 1.2. Let k be a field of characteristic zero, let A be a commutative k-algebra, let G be a finite group, and let $C = C^{\cdot}$ be a complex of $A[G]$-modules. (I.e. C is a complex of A-modules on which G acts by chain homomorphisms.) Denote by $C^G \subset C$ the subcomplex of G-invariants. Then the natural A-module homomorphism

$$H^i(C^G) \longrightarrow H^i(C)^G$$

is an isomorphism.

Proof. As A is an algebra over a field of characteristic zero, one can average over G to see that the left exact functor $M \longrightarrow M^G$ on $A[G]$-modules is right exact. Since evidently $B^i(C^G) = B^i(C)^G$ and $Z^i(C^G) = Z^i(C)^G$, taking invariants in the exact sequence $0 \longrightarrow B^i(C) \longrightarrow Z^i(C) \longrightarrow H^i(C) \longrightarrow 0$ yields the required isomorphism. ∎

We will apply Proposition 1.1 in the following relative setting:

Corollary 1.3. Let $f : X \longrightarrow Y$ be a Galois covering of complex algebraic varieties, with group G. Consider the commutative diagram

$$\begin{array}{ccc} X \times S & \xrightarrow{f \times 1} & Y \times S \\ & {\scriptstyle a} \searrow \quad \swarrow {\scriptstyle b} & \\ & S & \end{array}$$

<u>where</u> S <u>is some (complex algebraic) variety and</u> a <u>and</u> b <u>are the projections on the first factors. If</u> L <u>is a line bundle on</u> $Y \times S$, <u>then one has a canonical isomorphism</u>

$$R^i b_*(L) = R^i a_*((f \times 1)^* L)^G,$$

<u>where the term on the right is the</u> G - <u>invariant subsheaf of the</u> G-<u>sheaf</u> $R^i a_*((f \times 1)^* L)$.

<u>Proof</u>. The assertion is local on S, so we may suppose that $S = \mathrm{Spec}(A)$ is affine. Then $R^i b_*(L)$ is the sheaf associated to the A - module $H^i(Y \times S, L)$ and similarly for $R^i a_*((f \times 1)^* L)$. Therefore the Corollary follows by applying Proposition 1.1 to the Galois covering $f \times 1 : X \times S \longrightarrow Y \times S$. ∎

§2. The Computation

We keep notation as in the Introduction: thus C is a compact connected Riemann surface of genus $g \geq 2$, and Q is the vector bundle occurring in the statement of the Theorem.

<u>Lemma 2.1</u>. (Compare [K1], p. 326].) <u>Let</u> $\Delta \subset C \times C$ <u>be the diagonal, and denote by</u> $p : C \times C \longrightarrow C$ <u>the projection onto the first factor. Then</u>

$$R^0 p_* \mathcal{O}_{C \times C}(\Delta) = \mathcal{O}_C \quad \text{and} \quad R^1 p_* \mathcal{O}_{C \times C}(\Delta) = Q.$$

<u>Proof</u>. This follows from the exact sequence $0 \longrightarrow \mathcal{O}_{C \times C} \longrightarrow \mathcal{O}_{C \times C}(\Delta) \longrightarrow \mathcal{O}_C \longrightarrow 0$ upon taking direct images. ∎

Consider now the universal divisor $D_m \subset C \times C_m$ of degree m, so that $D_m \simeq C \times C_{m-1}$, and let $p : C \times C_m \longrightarrow C$ be projection onto the first factor.

<u>Proposition 2.2</u>. <u>One has a canonical isomorphism</u>

$$R^k p_*(\mathcal{O}_{C \times C_m}(D_m)) = \Lambda^k Q.$$

<u>Proof</u> The idea is to apply Corollary 1.3. To this end, denote by $C^m = C \times \ldots \times C$ the m-fold Cartesian product of C, and write $r : C^m \longrightarrow C_m$ for the canonical map, so that r is a Galois covering, with group the symmetric group S_m on m objects. Then one has a commutative diagram

$$\begin{array}{ccc} C \times C^m & \xrightarrow{1 \times r} & C \times C_m \\ & {\scriptstyle q}\searrow \quad \swarrow{\scriptstyle p} & \\ & C & \end{array}$$

where p and q are projections to C. Corollary 1.3 yields first of all:

$$R^k p_*(\mathcal{O}_{C \times C_m}(D_m)) = R^k q_*((1 \times r)^* \mathcal{O}_{C \times C_m}(D_m))^{S_m}. \tag{2.3}$$

To explicate the sheaf on the right, consider the maps

$$\pi_i = 1 \times pr_i : C \times C^m \longrightarrow C \times C,$$

pr_i being projection onto the ith factor. Then

$$(2.4) \qquad (1 \times r)^* \mathcal{O}_{C\times C_m}(D_m) = \bigotimes_{i=1}^{m} \pi_i^* \mathcal{O}_{C\times C}(\Delta) .$$

Set

$$R^0 = \mathcal{O}_C, \quad R^1 = Q, \quad \text{and} \quad R^j = 0 \quad \text{for } j \neq 0 \text{ or } 1,$$

so that the R^j are the sheaves appearing in Lemma 2.1. Since the direct images R^j are locally free, the Kunneth formula [EGA III.6.7.8] applies to (2.4) to give

$$(2.5) \qquad R^k q_*((1\times r)^* \mathcal{O}_{C\times C_m}(D_m)) = \bigoplus_{j_1+\ldots+j_m=k} (\bigotimes_{i=1}^{m} R^{j_i}).$$

Let F denote the direct sum of sheaves appearing on the right in (2.5). Each summand of F is naturally isomorphic to $T^k(R^1) \otimes T^{m-k}(R^0)$; let $\delta : T^k(R^1) \otimes T^{m-k}(R^0) \longrightarrow F$ denote the diagonal map. The action of S_m on F is such that $\sigma \in S_m$ carries the summand $R^{j_1} \otimes \ldots \otimes R^{j_m}$ of F to $R^{j_{\sigma(1)}} \otimes \ldots \otimes R^{j_{\sigma(m)}}$. Hence the S_m invariant subsheaf F^{S_m} of F is the image under δ of $T^k(R^1)^{S_k} \otimes T^{m-k}(R^0)^{S_{m-k}}$. Now $R^0 = \mathcal{O}_C$, so we may identify each summand of F with $T^k(R^1)$, and then $F^{S_m} = T^k(R^1)^{S_k}$. But the resulting diagonal map

$$\delta : T^k(R^1) = R^1 \otimes \ldots \otimes R^1 \longrightarrow F = R^k q_* \bigotimes_{i=1}^{m} \pi_i^* \mathcal{O}_{C\times C}(\Delta)$$

is alternating, since it is given by a sum of fibre-wise cup products. Hence

$$F^{S_m} = \Lambda^k R^1 = \Lambda^k Q,$$

proving the Proposition. ∎

Proof of the Theorem. As above, let $D_m \subset C \times C_m$ be the universal divisor of degree m, so that $D_m \simeq C \times C_{m-1}$ via the map

$$\begin{aligned} \gamma : C \times C_{m-1} &\longrightarrow C \times C_m \\ (P, D) &\longmapsto (P, D+P) . \end{aligned}$$

Let $f = f_m : C \times C_m \longrightarrow C_m$ denote projection onto the second factor. Then it is well known (c.f. [K1]) that

$$\Theta_{C_m} = f_* N$$

where $N = N_{D_m/C\times C_m}$ is the normal bundle to D_m in $C \times C_m$. Furthermore, the sheaf homomorphism du is identified with the connecting map

$$f_* N \longrightarrow R^1 f_* \mathcal{O}_{C\times C_m} = H^1(C, \mathcal{O}_C) \otimes \mathcal{O}_C$$

determined by the exact sequence

$$(2.6) \qquad 0 \longrightarrow \mathcal{O}_{C\times C_m} \longrightarrow \mathcal{O}_{C\times C_m}(D_m) \longrightarrow N \longrightarrow 0 .$$

But $f \mid D_m$ is finite, and therefore

$$(2.7) \qquad H^k(C_m, \Theta_{C_m}) = H^k(D_m, N) .$$

Moreover, $H^k(du)$ is identified with the composition

(2.8) $$H^k(D_m, N) \longrightarrow H^{k+1}(C \times C_m, \mathcal{O}_{C\times C_m}) \longrightarrow H^1(C, \mathcal{O}_C) \otimes H^k(C_m, \mathcal{O}_{C_m})$$

arising from (2.6) and the Kunneth decomposition.

The next step is to push the exact sequence (2.6) down to C. To this end, we use from [K1, p. 321] the fact that

$$\gamma^* N = p^*\mathcal{O}_C \otimes f^*\mathcal{O}_{C_{m-1}}(D_{m-1}),$$

where (somewhat abusively) we are writing p and f for the projections of $C \times C_{m-1}$ onto its first and second factors. Recalling that $H^k(C_m, \mathcal{O}_{C_m}) = \Lambda^k H^1(C, \mathcal{O}_C)$, and applying Proposition 2.2, the push-forward of (2.6) becomes

(2.9) $$\begin{array}{ccccc} R^{k-1}p_*N & \longrightarrow & R^k p_* \mathcal{O}_{C\times C_m} & \longrightarrow & R^k p_* \mathcal{O}_{C\times C_m}(D_m) \\ \| & & \| & & \| \\ \mathcal{O}_C \otimes \Lambda^{k-1}Q & & \Lambda^k H^1(C, \mathcal{O}_C) \otimes \mathcal{O}_C & & \Lambda^k Q \end{array}$$

The reader may check that under the indicated identifications, this is nothing but the exact sequence $(*)_k$ from the Introduction. [One analyzes the map $R^k p_* \mathcal{O}_{C\times C_m} \longrightarrow R^k p_* \mathcal{O}_{C\times C_m}(D_m)$ using the argument of Proposition 2.2.] In particular, (2.9) is a short exact sequence.

As $g(C) \geq 2$, one easily verifies that $H^0(C, \mathcal{O}_C \otimes \Lambda^k Q) = 0$ provided that $k < g-1$ (c.f. [GL2] or [L, §1]). Therefore, by the Leray spectral sequence:

$$H^k(D_m, N) = H^1(C, \mathcal{O}_C \otimes \Lambda^{k-1}Q)$$

when $k < g-1$, which proves the first statement of the Theorem. The analysis of $H^k(du)$ follows from (2.8) and (2.9). ∎

§3. Variants

In conclusion, we state some variants of the the theorem and its corollary.

To begin with, fix an integer $m \leq g-1$, and consider the set

$$C_m \times C_{2g-2-m} \supset Z_m = \{ (D, E) \mid D + E \in |K| \},$$

where K denotes a canonical divisor on C. There is a tautologous branched covering

$$f = f_m : Z_m \longrightarrow |K| = \mathbb{P}^{g-1}$$

which takes the pair of divisors (D, E) to the canonical divisor $D+E$. One may think of Z_m as the scheme parametrizing all possible ways of writing a canonical divisor as a sum of two effective divisors, of degrees m and $2g-2-m$ respectively. Let $L = f^*\mathcal{O}_{\mathbb{P}^{g-1}}(1)$ be the line bundle on Z_m defining the covering.

Proposition 3.1. Fix an integer $k \leq m-1$, and assume that property (N_{k-1}) holds for the canonical bundle Ω. Then (N_k) is satisfied if and only if

$$\dim H^k(Z_m, L) = g \cdot \binom{g}{k} + \binom{g-1}{k-1} \cdot (2k+1-3g)$$

Proof. Let N denote the cokernel of the sheaf homomorphism du, so that one has an exact sequence

$$0 \longrightarrow \Theta_{C_m} \xrightarrow{du} H^1(C, \Theta_C) \otimes \Theta_{C_m} \longrightarrow N \longrightarrow 0$$

of sheaves on C_m. Then $Z_m = \mathbb{P}(N)$, and the map $f : Z_m \longrightarrow \mathbb{P}H^1(C, \Theta)$ arises in the evident way from this sequence. We claim that:

$$(*) \qquad R^i\pi_* \Theta_{\mathbb{P}(N)}(1) = 0 \quad \text{for} \quad i \geq 1,$$

where $\pi : \mathbb{P}H^1(C, \Theta_C) \times C_m \longrightarrow C_m$ is the projection. In fact, write $\mathbb{P} = \mathbb{P}H^1(C, \Theta_C)$. Then $\mathbb{P}(N) \subset \mathbb{P} \times C_m$ is defined scheme-theoretically by the vanishing of the natural map

$$(**) \qquad \pi^*\Theta_{C_m} \longrightarrow f^*\Theta_{\mathbb{P}}(1).$$

Since $Z_m = \mathbb{P}(N) \longrightarrow \mathbb{P}$ is evidently finite, every component of $\mathbb{P}(N)$ has dimension $\leq g-1$. Therefore (**) exhibits $\mathbb{P}(N) \subset \mathbb{P} \times C_m$ as a local complete intersection, and in particular the Koszul complex determined by (**) is exact. The assertion (*) then follows by chasing through that complex.

It follows from (*) that $H^k(Z_m, L) = H^k(C_m, N)$ for all k. By the hypothesis and the main theorem, one has an exact sequence

$$0 \longrightarrow H^k(\Theta_{C_m}) \longrightarrow H^1(\Theta_C) \otimes H^k(\Theta_{C_m}) \longrightarrow H^k(N) \longrightarrow H^{k+1}(\Theta_{C_m}) \longrightarrow H^1(\Theta_C) \otimes H^{k+1}(\Theta_{C_m}).$$

Hence again invoking the theorem one finds that (N_k) holds for Ω if and only if

$$\dim H^k(X, N) = \dim H^1(\Theta_C) \otimes \Lambda^k H^1(\Theta_C) - \dim H^k(\Theta_{C_m}).$$

Recalling that $H^k(C_m, \Theta_{C_m}) = H^1(C, \Lambda^{k-1}Q \otimes \Theta_C)$, the assertion now follows with a computation. ∎

Finally, when $m = g-1$, the theorem ties up in an amusing way with the geometry of the theta divisor on $J(C)$, and in particular with some of the ideas used by Green in his analysis [G2] of quadrics of rank four containing the canonical curve. We follow the notation of [ACGH, Chapter VI, §4]. Assume henceforth that C is non-hyperelliptic, and let $D \subset C_{g-1}$ be the locus over which the Abel-Jacobi map fails to be finite. Denote by L the pull-back to C_{g-1} the pull-back of the principal polarization on $J(C)$, so that $L = K_{C_{g-1}}$. Green noted that one can view the second derivatives $\partial^2\theta/\partial z_i \partial z_j$ of the Riemann theta function as sections of $L \mid D$, thereby defining a map

$$f : \mathrm{Sym}^2 H^1(C, \Theta_C) \longrightarrow H^0(C_{g-1}, L \otimes \Theta_D).$$

Corollary 3.2. If C is non-hyperelliptic, then f fails to be surjective if and only if C is Petri-exceptional, i.e (N_1) fails for C.

Sketch of Proof. One has an exact sequence

$$0 \longrightarrow \Theta_{C_{g-1}} \longrightarrow u^* \Theta_{J(C)} \longrightarrow I_D \otimes L \longrightarrow 0$$

where I_D denotes the ideal sheaf of D in C_{g-1}, and by the Theorem (N_1) holds for C if and only if the map

$$H^1(u^* \Theta_{J(C)}) = H^1(C, \Theta_C) \otimes H^1(C, \Theta_C) \longrightarrow H^1(C_{g-1}, I_D \otimes L)$$

is surjective. But referring to diagram (4.4) on p. 258 of [ACGH], one sees that this is equivalent to the surjectivity of f. ∎

Remark. We suspect that the Corollary generalizes as follows. Set

$$V^k = \ker\{ H^1(C, \Theta_C) \otimes \Lambda^k H^1(C, \Theta_C) \longrightarrow \Lambda^{k+1} H^1(C, \Theta_C) \}.$$

Then presumably there is a map $f_k \colon V^k \longrightarrow H^{k-1}(C_{g-1}, \Theta_D \otimes L)$, and if (N_{k-1}) holds, then (N_k) should if and only if f_k is surjective.

References.

[ACGH] M. Arbarello, M. Cornalba, P. Griffiths and J. Harris, Geometry of Algebraic Curves (Springer Verlag), 1985.

[G1] M. Green, Koszul cohomology and the geometry of projective varieties, J. Diff. Geom. 19 (1984), pp 125-171.

[G2] M. Green, Quadrics of rank four in the ideal of a canonical curve, Inv. Math. 75 (1984), pp. 85-104.

[GL1] M. Green and R. Lazarsfeld, On the projective normality of complete linear series on an algebraic curve, Inv, Math. 83 (1986), pp. 73-90.

[GL2] M. Green and R. Lazarsfeld, A simple proof of Petri's theorem on canonical curves, in Geometry Today, , Progress in Math. Vol. 60, Birkhauser (1985), pp 129-142.

[Tohuku] A. Grothendieck, Sur quelques points d'algèbre homologique, Tohoku Math. J 9 (1957), pp. 119-221.

[EGA] A. Grothendieck and J. Dieudonné, Eléments de géometrie algébrique, III, Publ. Math. IHES 17 (1963).

[K1] G. Kempf, Deformations of symmetric products, in Riemann Surfaces and Related Topics, Annals of Math Studies No. 97, (1980), pp. 319-341.

[K2]. G. Kempf, Towards the inversion of abelian integrals, I, Ann. Math 110 (1979), pp. 243-273.

[L1] R. Lazarsfeld, A sampling of vector bundle techniques in the study of linear series, in M. Cornalba et. al (eds), Lectures on Riemann Surfaces (World Scientific Press, 1989), pp.500-559.

[L2] R. Lazarsfeld, Linear series on algebraic varieties, to appear in Proc. ICM Kyoto 1990.

[Muk] S. Mukai, Duality between D(X) and D($\hat{X}$) with an application to Picard sheaves, Nagoya Math J. 81 (1981), pp. 153 - 175.

[PR]. K. Paranjape and S. Ramanan, On the canonical ring of a curve, in Algebraic Geometry and Commutative Algebra, Academic Press, pp. 503-516.

10

Self-Dual Manifolds and Hyperbolic Geometry

Claude LeBrun*
SUNY Stony Brook
Stony Brook, New York

Abstract

An oriented Riemannian 4-manifold is said to be *half-conformally-flat* if its conformal curvature W, considered as a bundle-valued 2-form, is either self-dual or anti-self-dual. We describe a construction [29] of compact half-conformally-flat manifolds with semi-free isometric S^1-action, starting from the Green's function of a collection of points in a hyperbolic 3-manifold. It is then shown that any compact half-conformally-flat manifold with non-negative scalar curvature and semi-free isometric S^1-action either arises from this construction or is conformally flat. An application is then given to the existence and uniqueness of critical Kähler metrics in the sense of Calabi.

1 Introduction

One of the most natural and fundamental problems in Riemannian geometry is that of finding a "canonical" metric or class of metrics on each compact smooth manifold, the goal being, in part, to use geometry in order to better understand the topology of manifolds. While "canonical" may still mean different things to different people, flat metrics are by general agreement about as canonical as one could hope for, and most attempts at giving substance to the problem therefore involve minimizing some norm of the curvature over the space of metrics on the given manifold.

In dimension 2, the classical uniformization theorem allows one to find metrics of constant curvature in each conformal class, and this fact now plays a dominant role in most current research on Riemann surfaces. In dimension 3, Thurston's geometrization program [42] has come to play an analogous role in 3-dimensional topology. In higher dimensions, however, the situation remains comparatively unexplored.

*Supported in part by NSF grant DMS-9003263.

One of the most natural of such problems in dimension 4 is to find extrema (or perhaps just critical points) of the scale-invariant functional

$$\mathcal{K}(g) = \frac{1}{4\pi^2}\int_M \|\mathcal{R}_g\|^2 d\mathrm{vol}_g$$

over the space of smooth Riemannian metrics on a given smooth, compact, oriented 4-manifold M; here $\mathcal{R}_g$ denotes the Riemann curvature tensor of the Riemannian metric g. A major source of insight into the problem stems from the observation that this L^2-norm of the curvature tensor can be simplified by applying the generalized Chern-Weil formulas for the signature and Euler characteristic of M. These are respectively given by

$$\begin{aligned} \tau(M) &= \frac{1}{12\pi^2}\int_M (\|W_+\|^2 - \|W_-\|^2) d\mathrm{vol}_g \\ \text{and} \qquad \chi(M) &= \frac{1}{8\pi^2}\int_M (\|W_+\|^2 + \|W_-\|^2 + \frac{s^2}{24} - \frac{1}{6}\|\Phi\|^2) d\mathrm{vol}_g \end{aligned}$$

where s, Φ, W_+ and W_- respectively denote the scalar curvature, trace-free Ricci curvature, self-dual Weyl curvature, and anti-self-dual Weyl curvature[1] of an arbitrary Riemannian metric g on M. On the other hand, we can rewrite $\mathcal{K}(g)$ as

$$\mathcal{K}(g) = \frac{1}{4\pi^2}\int_M (\|W_+\|^2 + \|W_-\|^2 + \frac{s^2}{24} + \frac{1}{6}\|\Phi\|^2) d\mathrm{vol}_g ,$$

so that

$$\begin{aligned} \mathcal{K}(g) &= \frac{1}{12\pi^2}\int_M \|\Phi\|^2 d\mathrm{vol}_g + 2\chi(M) \\ &= \frac{1}{\pi^2}\int_M (\|W_\pm\|^2 + \frac{s^2}{48}) d\mathrm{vol}_g - 2(\chi(M) \pm 3\tau(M)) . \end{aligned}$$

Thus, we can s easily read off some sufficient conditions for a metric to be an absolute minimum of $\mathcal{K}$; namely, any of the following sets of conditions suffices:

1. $\Phi = 0$, i.e. g is an Einstein metric; or

---

[1]These are the pieces of the curvature tensor corresponding to the decomposition of the Young-tableau representation ⊞ of $GL(4, \mathbf{R})$ as a representation of $SO(4)$. Concretely, the Ricci and scalar curvatures are obtained by taking successive traces of $\mathcal{R}$. The trace-free Ricci curvature Φ and the the Weyl curvature W are then obtained by projecting to the kernels of these successive trace maps. The Weyl tensor then further decomposes into the two tensors $W_\pm := \frac{1}{2}(W \pm \star W)$, where the Hodge-star operator $\star$ treats W as a bundle-valued 2-form. The importance of the Weyl curvature stems primarily from the fact that it is conformally invariant, playing much the same rôle in conformal Riemannian geometry as is played by the Riemann curvature $\mathcal{R}$ in ordinary Riemannian geometry.

2. $W_- = 0$ and $s = 0$, i.e. g is a scalar-flat *self-dual* metric; or

3. $W_+ = 0$ and $s = 0$, i.e. g is a scalar-flat *anti-self-dual* metric.

One of our primary goals in this paper will be the complete classification of all compact scalar-flat anti-self-dual Riemannian 4-manifolds (M, g) *admitting non-trivial Killing fields.* Our main classification results in this direction will be described in §5. Notice that reversing the orientation of M interchanges the conditions $W_+ = 0$ and $W_- = 0$, so these results apply equally well to the scalar-flat self-dual case occurring above.

Related to the minimization problem with which we began our discussion is a minimization problem for *conformal* Riemannian metrics. Here a conformal Riemannian metric means the equivalence class

$$[g] = \{\alpha \cdot g \mid \alpha : M \stackrel{C^\infty}{\rightarrow} \mathbf{R}^+\}$$

determined by some Riemannian metric g. The functional

$$\begin{aligned} \mathcal{A}(g) &= \frac{1}{12\pi^2}\int_M \|W\|^2 d\text{vol}_g \\ &= \frac{1}{12\pi^2}\int_M (\|W_+\|^2 + \|W_-\|^2) d\text{vol}_g \end{aligned}$$

is conformally invariant, in the sense that $\mathcal{A}(g) = \mathcal{A}(\alpha \cdot g)\ \forall \alpha : M \stackrel{C^\infty}{\rightarrow} \mathbf{R}^+$. We immediately observe that $\mathcal{A}(g) \geq \pm\tau$ with equality iff $W_\mp = 0$, so that self-dual or anti-self-dual conformal metrics provide us with absolute minima of $\mathcal{A}$. In order to avoid distinguishing between these two cases, which differ only in orientation, we will call oriented Riemannian 4-manifold *half-conformally-flat* if either $W_+ = 0$ or $W_- = 0$.

The two simplest examples of compact self-dual manifolds are provided by the Riemannian symmetric spaces S^4 and $\mathbf{CP}_2$, oriented in the usual manner. Both of these manifolds admit isometric S^1-actions which are *semi-free*, meaning that the isotropy of any point is either trivial or all of S^1. In this article, we will describe a construction of all compact half-conformally-flat 4-manifolds which admit such actions and have non-negative scalar curvature.

Which smooth compact 4-manifolds M admit half-conformally-flat metrics? Certainly not all; for example, neither $S^2 \times S^2$ nor $\mathbf{CP}_2 \# \overline{\mathbf{CP}}_2$ admit such metrics, since these manifolds have signature zero, which would force any putative half-conformally-flat metric to be conformally-flat— whereas [25] the only simply-connected conformally-flat manifold is S^4! If one imposes extra conditions on the scalar curvature, moreover, the number of obstructions increases [26][28]. In particular, in the scalar-flat case of our original problem, inspection of the Chern-Weil formulae immediately yields [26] the Lafontaine inequality

$$W_\pm = 0, s = 0 \Rightarrow 2\chi \mp 3\tau \leq 0 \ ,$$

which is a sort of mirror image of the Hitchin-Thorpe inequality

$$\Phi = 0 \Rightarrow 2\chi \mp 3\tau \geq 0$$

for Einstein 4-manifolds. For this and other already-mentioned reasons, scalar-flat half-conformally-flat metrics have many of the more compelling features of Einstein metrics. However, the class of compact 4-manifolds which can admit metrics with $W_+ = 0$, $s = 0$, is very narrow indeed; either $b_+ = 0$ or any such a metric must be Kähler. Oddly enough, then, our primary problem in Riemannian geometry is largely a problem about extremal *Kähler* metrics in the sense of Calabi [6]. Our techniques for resolving much of the problem, however, will stem primarily from the conformal aspects of the problem rather than the Kählerian ones.

2 Twistors, Minitwistors, and Self-Duality

Perhaps the most compelling reason for the study of half-conformally-flat 4-manifolds is given by the the Penrose twistor correspondence [2] [37]. Let (M, g) be an orientable Riemannian 4-manifold, and let $F \to M$ be the principal $SO(4)$-bundle of orthonormal frames determining some orientation on M. Let $Z = F/U(2)$, which is a bundle over M with typical fiber $S^2 = SO(4)/U(2)$. Then the smooth 6-manifold Z carries a natural almost-complex structure $J : TZ \to TZ$, $J^2 = -1$, which leaves invariant both the tangent spaces of each fiber and the horizontal spaces of the metric connection of g. Indeed, let us notice that, by construction, Z is exactly the space of almost-complex structures $\jmath : TM \to TM$ compatible with the given metric and orientation, and so, thinking of the g-horizontal subspace of TZ as the pull-back of TM to Z, there is thus a tautological way to let J act on the horizontal sub-bundle of TZ. In the vertical directions, on the other hand, J will simply act as the standard complex structure on S^2, namely rotation by $+90°$. Provided that we give the fibers the correct orientation in defining this almost-complex structure J, the entire construction turns out, rather surprisingly, to be *conformally invariant*, meaning that J remains completely unchanged if the given Riemannian metric g is replaced by αg, where $\alpha : M \overset{C^\infty}{\to} \mathbf{R}^+$ is any smooth positive function. This construction of an almost-complex manifold for each conformal Riemannian manifold may thus be thought of as a higher-dimensional analogue of the correspondence between conformal Riemannian 2-manifolds and complex 1-manifolds. However, the almost-complex manifold (Z, J) will not, in general be a complex manifold— there need not be an atlas of charts for Z relative to which J identically becomes multiplication by i in $\mathbf{C}^3 = \mathbf{R}^6$. Instead, the relevant integrability condition turns out to either be $W_+ = 0$ or $W_- = 0$, depending on which orientation was chosen in constructing Z. In short, every half-conformally-flat 4-manifold determines a complex 3-fold Z, called its *twistor space*. For example, the twistor space of $\mathbf{S}^4$ is $\mathbf{CP}_3$, whereas the twistor space of $\mathbf{CP}_2$ is the flag-manifold

$$\mathsf{F} = \{([z_1, z_2, z_3], [w_1, w_2, w_3]) \in \mathbf{CP}_2 \times \mathbf{CP}_2 | z \cdot w = 0\} .$$

The class of complex 3-manifolds which arise by the twistor construction can be completely characterized in terms of their complex geometry, as follows. First, one needs a free anti-holomorphic involution $\sigma : Z \overset{\overline{\mathcal{O}}}{\to} Z$, $\sigma^2 = id_Z$, since such a map would arise in the above construction as the fiber-wise antipodal map; we will henceforth call such an involution a *real structure*. Secondly, there should exist at least one σ-invariant rational curve $\mathbf{CP}_1 \subset Z$ with normal bundle $\mathcal{O}(1) \oplus \mathcal{O}(1)$; in the above construction, any fiber of $M \to Z$ would fit the bill. A curve of the latter type is called a *real twistor line*, and it is not difficult to show that the real twistor lines locally foliate Z, and the 4-manifold M of such real twistor lines comes equipped with a canonical self-dual conformal metric for which the complex null vectors are precisely the images of those holomorphic sections of the the normal bundle of a leaf which have a zero somewhere. This provides an inverse to the above twistor construction.

Because the twistor construction is canonical, any conformal isometry of a self-dual 4-manifold induces a biholomorphism of its twistor space. By the same token, any conformal Killing field on M lifts to Z as the real part of a holomorphic vector field. If we have a free isometric S^1-action on M, we can therefore realize the corresponding twistor space as an open set in a holomorphic principal line bundle over a complex surface $\mathcal{T}$. The twistor lines will project to $\mathbf{CP}_1$'s with normal bundle $\mathcal{O}(2)$ in this surface, and the real structure will descend to a real structure τ on $\mathcal{T}$. The space $\mathcal{T}$ is called the *minitwistor space* of M/S^1, and the rational curves in $\mathcal{T}$ of self-intersection 2 which are invariant under the real structure are called *real minitwistor lines.* The space of real minitwistor lines is a real-analytic 3-manifold which may be identified with M/S^1; from its construction from $\mathcal{T}$, it is automatically endowed [20] with an *Einstein Weyl structure* for which $\mathcal{T}$ becomes the space of oriented geodesics. We will now describe this relationship between structures in dimensions 3 and 4 in in elementary differential geometric terms; for more details, cf. [22].

Suppose that (M, g) is a self-dual 4-manifold with a free isometric circle action generated by a Killing field ξ, and let $\xi^\flat = g(\xi, \cdot)$ be the corresponding 1-form. We may then equip the 3-manifold $X := M/S^1$ with the unique metric h for which the canonical projection π becomes a Riemannian submersion; i.e. such that

$$\pi^* h = (g - \frac{\xi^\flat \otimes \xi^\flat}{\|\xi\|^2}) .$$

Let ν be the unique 1-form on X such that

$$\pi^* \nu = \frac{-d\|\xi\|^2 + 2 \star \xi^\flat \wedge d\xi^\flat}{2\|\xi\|^2} , \tag{1}$$

and define a connection $\mathbf{D}$ on X by

$$\mathbf{D}_v w := \nabla_v w + \nu(v)w + \nu(w)v - g(v,w)\nu^{\#}, \tag{2}$$

where $\nu = h(\nu^{\#}, \cdot)$ and ∇ is the Riemannian connection of h. If we replace g by αg, where α is any S^1-invariant function on M, then the connection $\mathbf{D}$ and the conformal class $[h]$ remain unchanged. By construction, the torsion-free connection $\mathbf{D}$ preserves the conformal structure $[h]$ in the sense that parallel transport preserves angles, and is thus a so-called *Weyl connection*:

$$\mathbf{D}h = -2\nu \otimes h \ . \tag{3}$$

The hypothesis that (M, g) is self-dual then has the consequence that the symmetrization of the Ricci tensor $\mathbf{r}_{\mathbf{D}}$ of $\mathbf{D}$ is a multiple of h; i.e. there is a function $\lambda : X \to \mathbf{R}$ such that

$$\mathbf{r}_{\mathbf{D}}(v,w) + \mathbf{r}_{\mathbf{D}}(w,v) = \lambda h(v,w) \quad \forall\, v, w \ . \tag{4}$$

We will call a 3-manifold X equipped with a conformal metric h and a connection $\mathbf{D}$ satisfying (3) and (4) an *Einstein-Weyl manifold.* Such 3-dimensional geometries were first studied by Elie Cartan [7].

In order to reconstruct a self-dual 4-manifold from an Einstein-Weyl geometry, we need an extra piece of information, namely the function $V = \|\xi\|^{-1}$. If we think of $M \to X$ as a circle bundle, we may equip it with a connection θ whose horizontal spaces are the g-orthogonal complements of the fibers. The self-duality of g then implies that the curvature of θ is given by

$$d\theta = \star(d - \nu)V \ . \tag{5}$$

We may invert this construction as follows: let $(X, [h], \mathbf{D})$ be an Einstein-Weyl 3-manifold, and let $V : X \to \mathbf{R}$ be a positive solution of the elliptic equation

$$d \star (d - \nu)V = 0 \ .$$

Assume, in addition, that the closed 2-form $\frac{1}{2\pi} \star (d - \nu)V$ represents an integral class in the deRham cohomology $H^2(X)$. Then, by the Chern-Weil theorem, there is a circle-bundle $\pi : M \to X$ which admits a connection θ whose curvature is $d\theta = \star(d - \nu)V$. Then, for any positive function μ on M, the metric

$$g = \mu(\pi^* h + V^{-2}\theta \otimes \theta)$$

is self-dual. Often one takes $\mu = V$, so that the above expression becomes

$$g = Vh + V^{-1}\theta \otimes \theta \ .$$

We now complete our circle of ideas by associating a complex surface to an Einstein-Weyl geometry. Assume for simplicity that $(X, \mathbf{D})$ is geodesically convex, and let $\mathcal{T}$ denote the space of directed geodesics of $\mathbf{D}$, which is then a smooth 4-manifold diffeomorphic to TS^2. The tangent space of $\mathcal{T}$ at a geodesic γ is then just the space of solutions of *Jacobi's equation*

$$\mathbf{D}_\xi \mathbf{D}_\xi \eta = \mathcal{R}_{\xi\eta}\xi \ ,$$

modulo fields tangent to γ; here ξ denotes a tangent field of γ satisfying $\mathbf{D}_\xi \xi = 0$ and $\mathcal{R}^a{}_{bcd}$ is the curvature of the torsion-free connection $\mathbf{D}$. For an Einstein-Weyl 3-manifold, this simplifies to become

$$\mathbf{D}_\xi \mathbf{D}_\xi \eta \equiv -\frac{\mathcal{R}}{6}(\xi \cdot \xi)\eta - 5(\xi \cdot v)(\xi \times \eta) \ \ \mathbf{mod}\ \xi \ ,$$

where $\mathcal{R} = h^{ab}\mathcal{R}_{ab}$ is the scalar curvature, and $v^a = \epsilon^{abc}\mathcal{R}_{bc}$ is the Hodge-star of the skew part of the Ricci tensor, and inner and cross-products are with respect to h. The solution space of this equation is then invariant under the 90° rotation $\eta \mapsto \frac{\xi \times \eta}{\|\xi\|}$, and this then gives $\mathcal{T}$ the structure of an almost-complex manifold. Moreover, this almost-complex structure is automatically integrable, as follows seeing this in the real-analytic case is to identify $\mathcal{T}$ with the space of totally geodesic null planes in a small complexification of $(X, [h], \mathbf{D})$, which is manifestly a complex surface; more generally, this integrability follows from that of the twistor CR structure on the sphere-bundle of a conformal 3-fold, since the latter induces the above almost-complex structure by projection. The 2-sphere of directed $\mathbf{D}$-geodesics through any point $p \in X$ now becomes a holomorphic curve $\mathcal{C} \subset \mathcal{T}$ of self-intersection 2, while the map $\sigma : \mathcal{T} \to \mathcal{T}$ obtained by reversing the direction of each geodesic becomes an anti-holomorphic involution of $\mathcal{T}$.

The complex surface $\mathcal{T}$ is the *minitwistor space* of the given EinsteinWeyl geometry $(X, [h], \mathbf{D})$; knowing this space together with the anti-holomorphic involution $\tau : \mathcal{T} \to \mathcal{T}$ allows us to completely reconstruct the original Einstein-Weyl space. Indeed, X is precisely the space of smooth, embedded, τ-invariant, compact holomorphic curves in $\mathcal{T}$ with self-intersection 2, and the family of such curves passing through a given point (and hence also passing through its τ-conjugate point) is a geodesic of the connection $\mathbf{D}$. Each point of such a curve $\mathcal{C}_p$ may now be thought of as representing a point in the sphere of directions $(T_pX - 0)/\mathbf{R}^+$ at the corresponding point $p \in X$, and the conformal structure of each such $\mathcal{C}_p$ therefore equips X with a conformal structure, namely $[h]$. But indeed, we actually could have started with *any* complex surface $\mathcal{T}$ equipped with a fixed-point-free antiholomorphic involution τ and containing a τ-invariant rational curve $\mathcal{C}$ of self-intersection 2; the above prescription would then construct the general Einstein-Weyl 3-fold, starting with this essentially holomorphic data.

Over such a complex surface $\mathcal{T}$, let us suppose we have a holomorphic line bundle L whose Chern class vanishes. Suppose, moreover, that the involution τ lifts to an anti-holomorphic involution σ of the total space of L. Via the Penrose transform, we may analyze L in three different ways:

- We may view L as the exponential of an element of $H^1(\mathcal{T}, \mathcal{O})$, at least in a neighborhood of the curve $\mathcal{C}$. Via the Penrose transform, this gives us a function V (secretly a section of the conformal weight -1 line bundle) satisfying
$$d\star(d-\nu)V = 0,$$
where $\star$ denotes the Hodge star operator of h.

- We may instead apply the Hitchin-Ward correspondence [21] directly to L to obtain a solution of the "$\mathbf{U(1)}$ Bogomolny equations"
$$d\theta = \star(d-\nu)V$$
on $(X, [h], \mathbf{D})$. Here $d\theta$ is the curvature of a complex line bundle over X, while the "Higgs field" V is precisely the same section of the conformal weight -1 line bundle encountered before.

- Finally, we may attempt to treat the complex manifold L as a twistor space, defining M to be the set of σ-invariant rational complex curves in L with normal bundle $\mathcal{O}(1)+\mathcal{O}(1)$. Any such curve projects to a τ-invariant rational curve in $\mathcal{T}$ of self-intersection 2. Conversely, for each τ-invariant rational curve $\mathcal{C}$ in $\mathcal{T}$ of self-intersection 2, there is a circle's worth of σ-invariant rational complex curves in L obtained as holomorphic sections of $C|_L$; the normal bundle of such a curve is *a priori* an extension
$$0 \to \mathcal{O} \to N \to \mathcal{O}(2) \to 0,$$
and we may identify the splitting obstruction $\in H^1(\mathbf{CP}_1, \mathcal{O}(-2))$ for this sequence with the value of V at the corresponding point of X. If $V \neq 0$ is real-valued, $L-0$ is therefore the twistor space of a half-conformally-flat 4-manifold. Indeed, this manifold is exactly [22] the circle bundle with curvature $d\theta = \star(d-\nu)V$, and the conformal metric is just
$$[g] = [Vh + V^{-1}\theta^2]\ .$$
Conversely, every half-conformally-flat 4-manifold with S^1-action arises in this way.

3 Examples

Example 1. Take X to be $\mathbf{R}^3$ punctured at n points $q_1, \ldots, q_n$, with h the Euclidean metric and $\mathbf{D}$ the usual flat connection. Let V be the sum of the Green's functions of the given points:
$$V = \sum_{j=1}^{n} \frac{1}{2r_j}\ ,$$

where r_j is the Euclidean distance from q_j. Then $\frac{1}{2\pi} \star dV$ has integral -1 on a small sphere around any one of the puncture points q_j; since such spheres generate $H_2(\mathbf{R}^3 - \{q_1, \dots, q_n\})$, we conclude that $[\frac{1}{2\pi} \star dV]$ is an integral deRham class. We can therefore consider the circle bundle $M_0 \to (\mathbf{R}^3 - \{q_1, \dots, q_n\})$ with connection 1-form θ whose curvature is $\star dV$. There is then a self-dual metric on M given by

$$g = Vh + V^{-1}\theta \otimes \theta \, .$$

This is the metric of Gibbons and Hawking [14]. If we add n points $p_1, \dots, p_n$ to M_0 to obtain a new space M which comes equipped with a circle action having $p_1, \dots, p_n$ as its fixed points and a projection $M \to \mathbf{R}^3$, then M admits a unique smooth structure such that g extends to M as a smooth Riemannian metric:

$$\begin{array}{ccccc} M & = & M_0 & \cup & \{p_1, \dots, p_n\} \\ \downarrow & & \downarrow & & \downarrow \\ \mathbf{R}^3 & = & (\mathbf{R}^3 - \{q_1, \dots, q_n\}) & \cup & \{q_1, \dots, q_n\} \end{array}$$

Moreover, the resulting Riemannian manifold is complete and, by virtue of special properties of the Einstein-Weyl space $\mathbf{R}^3$, actually Ricci-flat Kähler. With a little care, this construction can easily be generalized to the case of infinitely many (sparsely located) centers [1].

Example 2 [29]. Let $(X, h, \mathbf{D})$ be hyperbolic 3-space $\mathcal{H}^3$ punctured at n points $q_1, \dots, q_n$, where $\mathbf{D} = \nabla$ is the Riemannian connection. We again build V from the Green's functions of the given points:

$$V = 1 + \sum_{j=1}^{n} \frac{1}{e^{2r_j} - 1} \, ,$$

where r_j denotes the hyperbolic distance from q_j. Then $[\frac{1}{2\pi} \star dV]$ is again an integral class, and we can define a circle bundle $M_0 \to (\mathcal{H}^3 - \{q_1, \dots, q_n\})$ with connection 1-form θ whose curvature is $\star dV$. Let r denote the hyperbolic distance from any reference point. The metric

$$g = (\operatorname{sech}^2 r)\,(Vh + V^{-1}\theta^{\otimes 2}) \qquad (6)$$

is then self-dual, and, because of our choice of conformal gauge, may be smoothly compactified by adding a 2-sphere and n points $p_1, \dots, p_n$. Indeed, let B denote the closed unit ball in $\mathbf{R}^3$, and identify the interior of B with $\mathcal{H}^3$ via the Poincaré conformal model. Then $M = M_0 \cup S^2 \cup \{q_1, \dots, q_n\}$ can be made into a smooth 4-manifold with circle-action in such a manner that $S^2 \cup \{q_1, \dots, q_n\}$ is the fixed point set and B is the orbit space, so that the projection to B is as follows:

$$\begin{array}{ccccccc} M & = & M_0 & \cup & S^2 & \cup & \{p_1,\ldots,p_n\} \\ \downarrow & & \downarrow & & \downarrow & & \downarrow \\ B & = & (\mathcal{H}^3 - \{q_1,\ldots,q_n\}) & \cup & \partial B & \cup & \{q_1,\ldots,q_n\} \end{array}$$

Calculations similar to those involved in the analysis of *Example 1* then show

Theorem. *The metric g of equation (6) has non-negative scalar curvature, and extends to M to yield a compact self-dual 4-manifold diffeomorphic to the n-fold connected sum* $\mathbf{CP}_2\#\cdots\#\mathbf{CP}_2$.

When $n = 0, 1$, this construction produces the standard conformal metrics on $\mathbf{S}^4$ and $\mathbf{CP}_2$, respectively. When $n = 2$, we instead get the metrics of Poon [39].

Example 3 [30], [31], [23]. Let X be a hyperbolic 3-manifold with smooth conformal compactification X, meaning that we assume that Y is a smooth compact 3-manifold-with-boundary such that $X = Y - \partial Y$, and the hyperbolic metric h of X is of the form $h = f^{-2}\hat{h}$ for $\hat{h}$ a smooth Riemannian metric on Y and f a non-degenerate defining function of the boundary ∂Y; by Thurston's main theorem [42], the class of Y admitting structures of this kind includes "most" atoroidal 3-manifolds-with-boundary. Let $q_1,\ldots,q_n \in Y$ be given, let G_j be the corresponding Green's functions, and set $X = Y - \{q_1,\ldots,q_n\}$. Then we can mimic the previous construction of compact self-dual 4-manifolds by taking

$$V = 1 + \sum G_j ,$$

trying to find a circle-bundle with connection 1-form θ whose curvature is $\star dV$, setting

$$g = f^2 \left(\pi^* V h + V^{-1}\theta \otimes \theta\right) ,$$

and compactifying by adding a copy of ∂Y and n isolated fixed points $\{p_1,\ldots,p_n\}$. The only catch lies in showing that $[\frac{1}{2\pi} \star dV]$ is an integral cohomology class— and, indeed, this will usually only be true for some special configurations of points! Nonetheless, one can verify the integrality condition in many cases. For example, if X is a handle-body, the integrality condition is automatically verified, and one may use this to construct explicit self-dual metrics on arbitrary connected sums of $S^1 \times S^3$'s and $\mathbf{CP}_2$'s. On the other hand, if $Y = S_{\mathbf{g}} \times \mathbf{R}$, where $S_{\mathbf{g}}$ is a compact surface of genus $\mathbf{g} \geq 2$, one finds that the integrality condition is non-trivial, but, by restricting ones choice of point-configurations, the construction can be made to yield self-dual metrics on $(S^2 \times S_{\mathbf{g}})\# n\mathbf{CP}_2$ provided that $n \neq 1$.

All the self-dual manifolds described in this section are of course associated with twistor spaces, and these complex 3-manifolds completely encode each conformal geometry. The fact that the metrics in question have conformal Killing fields is reflected by a $\mathbf{C}_*$-action on their twistor spaces, and, at least locally, the twistor spaces in question therefore fiber over complex complex surfaces, called the *minitwistor space* of the relevant Einstein-Weyl quotient geometries, as explained in the last section. Let us now examine our examples in this light.

We begin with the Gibbons-Hawking metrics of *Example 1*, the twistor spaces of which were discovered by Hitchin [18]. The relevant Einstein-Weyl geometry is in this case that of Euclidean 3-space, and the corresponding mini-twistor space [21] is $T\mathbf{CP}_1$. Let $\mathcal{O}(k) \to T\mathbf{CP}_1$ denote the pull-back of the degree k line-bundle over $\mathbf{CP}_1$ via the canonical projection. The data points $q_1, \ldots, q_n \in \mathbf{R}^3$ specify n sections of $T\mathbf{CP}_1 \to \mathbf{CP}_1$, and these are the zero loci of n sections $P_1, \ldots, P_n$ of $\mathcal{O}(2)$. In the total space of the rank 2 vector bundle $\mathcal{O}(n) \oplus \mathcal{O}(n)$, let $\tilde{Z}$ denote the hypersurface

$$xy = \prod_{j=1}^{n} P_j ,$$

where x and y refer to the two factors of $\mathcal{O}(n) \oplus \mathcal{O}(n)$. The twistor space Z of the Gibbons-Hawking metric is then given by a "small resolution" of this 3-dimensional complex algebraic variety, meaning that each singular point is replaced by a rational curve. For an important generalization of this class of twistor spaces, see [24].

We now turn to the manifolds given by *Example 2*. In this case, the relevant Einstein-Weyl geometry is that of hyperbolic 3-space, and the corresponding minitwistor space is $\mathbf{CP}_1 \times \mathbf{CP}_1$. Let $\mathcal{O}(k, \ell)$ denote the unique holomorphic line-bundle over $\mathbf{CP}_1 \times \mathbf{CP}_1$ with degree k on the first factor and degree ℓ on the second, and let. The data points $q_1, \ldots, q_n \in \mathcal{H}^3$ correspond to the zero loci of n sections $P_1, \ldots, P_n \in \Gamma(\mathbf{CP}_1 \times \mathbf{CP}_1, \mathcal{O}(1,1))$. Let $\mathcal{B}$ denote the total space of the $\mathbf{CP}_2$-bundle

$$\mathcal{B} := \mathbf{P}(\mathcal{O}(n-1,1) \oplus \mathcal{O}(1,n-1) \oplus \mathcal{O}) \xrightarrow{\pi} \mathbf{CP}_1 \times \mathbf{CP}_1 ,$$

and define an algebraic variety $\tilde{Z} \subset \mathcal{B}$ by the equation

$$xy = t^2 \prod_{j=1}^{n} P_j ,$$

where $x \in \mathcal{O}(n-1,1)$, $y \in \mathcal{O}(1,n-1)$, and $t \in \mathcal{O} := \mathcal{O}(0,0)$. The twistor space Z of the metric constructed in *Example 2* is then obtained from $\tilde{Z}$ by making small resolutions of the singular points and blowing down the surfaces $x = t = 0$ and $y = t = 0$ to $\mathbf{CP}_1$'s. Notice that Hitchin's twistor spaces are degenerations of these.

The twistor spaces of these manifolds therefore turn out to be *Moishezon spaces*, meaning that they are bimeromorphic to smooth projective varieties.

However, one can also show [32] [9] that their generic small deformations are not bimeromorphic to any Kähler manifold, so that one observes from these examples that the class of compact complex manifolds bimeromorphic to Kähler is not stable under deformation of complex structure.

Finally, a word concerning *Example 3.* The space of geodesics of a general hyperbolic 3-manifold is not typically a Hausdorff spaces, so there is no reasonable *global* of the twistor spaces of these manifolds that would present them as families of curves over a complex surface. However, if one passes to the *universal cover* of such an example, the twistor space will indeed become a conic bundle over an open subset of $\mathbf{CP}_1 \times \mathbf{CP}_1$ once one blows up a pair of curves for each boundary component of X.

4 Classification Theorems

In this section, we shall describe sufficient conditions for a self-dual manifold to arise by the hyperbolic ansatz described in the previous section. We begin with some general remarks about circle actions on 4-manifolds and their fixed points.

Definition 1 *An non-trivial action of a group G on a connected space M is said to be* semi-free *if, for every $x \in M$, the corresponding isotropy subgroup $I_x < G$ of is either all of G or else is the trivial subgroup $\{1\}$.*

When S^1 acts semi-freely on a 4-manifold M, the quotient $Y = M/S^1$ is a compact 3-manifold-with-boundary, the boundary ∂Y of which is given by the set of non-isolated fixed points of the action. To see this, let us first observe that M must admit Riemannian metrics which are invariant under the action; such metrics can in fact be constructed by averaging any given metric over the action. Using the exponential map at any fixed point, the action in a normal neighborhood becomes differentiably conjugate to an linear orthogonal action of S^1 on Euclidean space $\mathbf{R}^4$, and so, relative to a suitable oriented orthonormal frame at the fixed-point, can be put in the normal form

$$e^{it} \mapsto \begin{bmatrix} \cos kt & \sin kt & 0 & 0 \\ -\sin kt & \cos kt & 0 & 0 \\ 0 & 0 & \cos \ell t & \sin \ell t \\ 0 & 0 & -\sin \ell t & \cos \ell t \end{bmatrix}$$

for integers k, ℓ. The semi-free condition then allows us to take $k = 1$ and $\ell \in \{0, \pm 1\}$. If $\ell = 0$, our fixed point, rather than being isolated, belongs to a smooth 2-dimensional surface of fixed points, henceforth called the *fixed surface* of the action; the quotient of M by S^1 can locally be identified with the upper half-space in $\mathbf{R}^3$, the quotient map being given by

$$(x^1, x^2, x^3, x^4) \mapsto (x^3, x^4, (x^1)^2 + (x^2)^2) ,$$

so that the fixed surface exactly becomes the set of boundary points in the quotient. If, on the other hand, $\ell = \pm 1$, so that the fixed point in question is isolated, the quotient can locally be identified with $\mathbf{R}^3$, the quotient map being given by

$$(x^1, x^2, x^3, x^4) \to (x^1x^3 \pm x^2x^4, x^2x^3 \mp x^1x^4, (x^1)^2 + (x^2)^2 - (x^3)^2 - (x^4)^2) \,.$$

For further background on circle actions on 4-manifolds, cf. e.g. [11][15].

Definition 2 *Let (M, g) be a compact oriented Riemannian 4-manifold. A smooth S^1-action on M with fixed surface Σ will be called* docile *if the following are satisfied:*

1. *the action is conformally isometric and semi-free;*
2. *there is a non-isolated fixed point; and*
3. *at any* isolated *fixed point $p \in M$, the induced action on $(\wedge_-)_p$ is trivial.*

Example Isometric rotation of S^4 about an equatorial S^2, as given by

$$\begin{bmatrix} \cos t & \sin t & 0 & 0 & 0 \\ -\sin t & \cos t & 0 & 0 & 0 \\ 0 & 0 & 1 & 0 & 0 \\ 0 & 0 & 0 & 1 & 0 \\ 0 & 0 & 0 & 0 & 1 \end{bmatrix}$$

is docile. By contrast, the semi-free action of S^1 on S^4 by rotation about 2 isolated fixed points, as given by

$$\begin{bmatrix} \cos t & \sin t & 0 & 0 & 0 \\ -\sin t & \cos t & 0 & 0 & 0 \\ 0 & 0 & \cos t & \sin t & 0 \\ 0 & 0 & -\sin t & \cos t & 0 \\ 0 & 0 & 0 & 0 & 1 \end{bmatrix}$$

fails to be docile for two distinct reasons: there are *only* isolated fixed points, and the action on $\wedge_-$ is non-trivial at one of the fixed points.

Docility might appear to be an overly-stringent condition. Remakably, however, it is automatic in many contexts, as we demonstrate in the next pair of propositions:

Proposition 1 *Suppose that M is a smooth, compact, oriented 4-manifold with strictly positive intersection form. Then any semi-free circle action on*

M is necessarily docile. Moreover, the quotient $Y = M/S^1$ is a smooth 3-manifold with vanishing second Betti number and non-empty, connected boundary.

Proof. Let Σ denote the fixed surface of the action. Because the action is semi-free, this agrees with the fixed surface of the involution corresponding to $-1 \in S^1$. By [3], the self-intersection the fixed point set of any involution homotopic to the identity must be the signature of M; that is, $[\Sigma]\cdot[\Sigma] = \tau > 0$, so in particular $\Sigma \neq \emptyset$, and there are non-isolated fixed points. Since the action is semi-free, the quotient $Y = M/S^1$ is a smooth 3-manifold with non-empty boundary $\partial Y \cong \Sigma$. Moreover, ∂Y must be connected, for otherwise a curve in Y joining two distinct components would have as inverse image in M a 2-sphere with self-intersection 0 and yet having non-trivial topological intersection numbers with two of the components of Σ; this would thus give rise to a non-trivial class in $H_2(M)$ with trivial self-intersection, contradicting the positivity of the intersection form.

We will now show that $H^2(Y, \mathbf{R}) = 0$. Let $[\omega] \in H^2_{\text{deRahm}}(Y)$ be given. By the Poincaré lemma, $[\omega]$ may be represented by a a smooth closed differential 2-form ω which vanishes in a neighborhood of the images of the isolated fixed points of the action. Let Σ denote the fixed surface of the action and let $\pi : M \to Y$ denote the canonical smooth projection. Since $\pi^*\omega \wedge \pi^*\omega = \pi^*[\omega \wedge \omega] = 0$, the fact that $\cup : H^2(M) \times H^2(M) \to \mathbf{R}$ is positive and $[\omega] \cup [\omega] = 0$ implies that $[\pi^*\omega] = 0$ in deRham cohomology. Hence $\omega = d\phi$ for some 1-form ϕ. Since ϕ is closed near the isolated fixed points, we may, by the Poincaré lemma, choose ϕ to vanish identically there, and, by an elementary line-integral argument, we may also choose ϕ to have vanishing radial component near the fixed surfaces. By replacing ϕ with its average over the action, we may now assume that ϕ is an *invariant form*, still vanishing in a neighborhood of the isolated fixed points and having vanishing radial component near the fixed surface Σ. Let ξ denote the vector field which generates the action. Then

$$0 = \mathcal{L}_\xi \phi = \xi \rfloor d\phi + d(\xi \rfloor \phi) = d(\xi \rfloor \phi)$$

so that $\xi \rfloor \phi$ is constant. But since this expression vanishes at the fixed points, $\xi \rfloor \phi \equiv 0$. Since our gauge-fixing has arranged for ϕ to be everywhere in the image of π^*, we have $\phi = \pi^*\psi$ for a smooth 1-form ψ on $\overline{Y}$. As $d\psi = \omega$, we conclude that $[\omega] = 0$ in deRham cohomology.

It remains to verify that condition (3) is satisfied. Recall that, in principle, our circle action might act near an isolated fixed-point p in one of two ways. By choosing a suitable orthonormal frame, the circle action on T_pM can be put in the form

$$\begin{bmatrix} \cos t & \sin t & 0 & 0 \\ -\sin t & \cos t & 0 & 0 \\ 0 & 0 & \cos t & \sin t \\ 0 & 0 & -\sin t & \cos t \end{bmatrix}$$

and the issue at hand is the *orientation* of such a basis; this is precisely equivalent to asking whether the action is trivial on $(\Lambda_+)_p$ or rather on $(\Lambda_-)_p$. To settle this, let q be the image of p in Y, and choose an embedded differentiable curve γ from q to ∂Y, avoiding the images of any other isolated fixed points. The inverse image in M of γ is a 2-sphere which joins q to Σ, and the self-intersection of this 2-sphere is ± 1, since we can can vary γ through curves γ_t from q to ∂Y which meet γ only at q. Indeed, the tangent spaces at p of the spheres corresponding to the γ_t will be complex lines in $\mathbf{C}^2 = \mathbf{R}^4$ under the identification of T_pM with $\mathbf{R}^4$ induced by the above orthonormal frame. The sign of the self-intersection is therefore $+$ if our diagonalizing frame determines the standard orientation, and $-$ if the it determines the opposite orientation. Since the intersection form is positive, it follows that the diagonalizing frame is oriented, and the action is trivial on $(\Lambda_-)_p \subset \Lambda^{1,1}\mathbf{C}^2$. ∎

Proposition 2 *Suppose that (M, g) is a compact self-dual 4-manifold with non-negative scalar curvature. Suppose that (M, g) is not conformally flat. Then any conformally isometric semi-free circle action on (M, g) must be docile.*

Proof. The Weitzenböck formula

$$(d + d^*)^2 = \Delta - 2W + \frac{s}{3}$$

for 2-forms implies [28] that any anti-self-dual harmonic 2-form on a self-dual 4-manifold with scalar curvature $s \geq 0$ must be parallel; one then concludes that either the intersection form $H^2 \times H^2 \to \mathbf{R}$ is positive, or else the manifold is reverse-oriented Kähler and scalar-flat. The former case is covered by Proposition 1. In the latter case, the Kähler form is harmonic, and this is a conformally invariant condition on middle-dimensional forms; the uniqueness clause of the Hodge theorem guarantees that the Kähler form must therefore be invariant under the group action, so that the action is holomorphic and isometric. We must now use this to show that, with respect to the anti-complex orientation, the action on Λ_- is trivial at each isolated fixed point.

Let ξ denote the Killing field which generates the given circle action, and let J denote the complex structure tensor. Since (M, g, J) is a compact Kähler manifold of constant scalar curvature, we can apply the theory of *holomorphy potentials* [33] to the Killing fields of M. To do so first notice that there are no parallel vector fields, since such a field would give rise to a local Riemannian splitting $M \sim \mathbf{C} \times S$, and the vanishing of the scalar curvature would then force the local factor S to be flat; this would in turn imply that M would be flat, contradicting the assumption that M is not conformally flat. Hence $J\xi = \text{grad} f$ for some smooth function f on M, namely the holomorphy potential of ξ. Notice that f is constant on the orbits of the action— in fact, ξ is the Hamiltonian vector field of f— and, because ξ is Killing, f is a non-degenerate Morse function in the sense of Bott [4], with relative extrema occurring at fixed surfaces and saddle points occurring at the isolated fixed points.

Since f is constant on the orbits of the action, we may choose to view f as a function $\check{f}$ on $Y = M/S^1$. Since f has non-degenerate critical points precisely at those points where the projection $\pi : M \to Y$ has vanishing derivative, and has Bott-non-degenerate critical points at the fixed surfaces of the action, where the normal derivative of π vanishes precisely to first order, $\check{f} : Y \to \mathbf{R}$ is a smooth function without critical points, and is constant on each component of ∂Y. We conclude that $Y \approx \Sigma_{\mathbf{g}} \times [-1, 1]$ for some compact oriented surface $\Sigma_{\mathbf{g}}$, and in particular the restriction map $H^2(Y, \mathbf{R}) \to H^2(\partial Y, \mathbf{R})$ is injective, so that condition (4) is satisfied. As conditions (1) and (2) are also trivially true, we now merely need to check condition (3).

Since $J\xi$ is gradient-like, any flow line of $J\xi$ must connect two fixed points of the action. Let $\Upsilon : [-1, 1] \to M$ be a curve whose endpoints are fixed points and whose interior reparameterizes a flow line of $J\xi$. We can then generate a holomorphic 2-sphere in (M, J) by sweeping Υ along by the S^1 action. To calculate the self-intersection of this holomorphic curve, we apply the same reasoning used in Proposition 1; namely, we vary the image γ of Υ in $Y = M/S^1$ while keeping the endpoints fixed. With respect to the original *anti-complex* orientation of M, the self-intersection is then given by $\iota_0 + \iota_1$, where the index $\iota_j \in \{0, \pm 1\}$ of the fixed point $\Upsilon(j)$ is 0 if $\Upsilon(j)$ is a non-isolated fixed point, and ± 1 if the action at the isolated fixed point is trivial on $\Lambda_{\mp}$. With respect to the complex orientation of (M, g) this then gives a rational curve of self-intersection $-\iota_0 - \iota_1$.

To finish the proof, we show there cannot be a fixed point of index -1. If there were, we could follow a chain of downward flow lines from it to a relative minimum of f, which would be a fixed point of index 0. If one flow line sufficed for such a journey, the rational curve corresponding to this flow line would have self-intersection $+1$. Otherwise, by deleting superfluous initial segments of such a trip, we may assume that all the intermediate stops are fixed points of index $+1$. Sweeping these flow lines along the S^1 action, we then produce a chain of rational curves with intersection matrix

$$\begin{bmatrix} 0 & 1 & & & \\ 1 & -2 & 1 & & \\ & 1 & \ddots & 1 & \\ & & 1 & -2 & 1 \\ & & & 1 & -1 \end{bmatrix}$$

with respect to the complex orientation. Successively blowing down (-1)-curves, we eventually produce a rational curve of self-intersection $+1$. The linear system of such a divisor gives a rational map to $\mathbf{CP}_2$, so our complex surface (M, J) must then be $\mathbf{CP}_2$ blown up at a collection of points. But (M, J) admits a holomorphic vector field $\xi - iJ\xi$. The blown-up points in $\mathbf{CP}_2$ must therefore be collinear. If the points in question are all distinct, such a blow-up of $\mathbf{CP}_2$ admits a non-zero holomorphic section of the anti-canonical line bundle κ^{-1} with divisor D given by $3L + 2\sum E_j$, where L is the proper transform of the line and the E_j are the exceptional curves obtained by blowing up the point in question; if some of the points coincide, say with maximal multiplicity m, a positive power κ^{-m} of the anti-canonical line bundle admits a holomorphic section, whose divisor we will again denote by D. Since the Ricci form P, the scalar curvature s, and the Kähler form Ω are related by

$$s\Omega \wedge \Omega = 4\mathrm{P} \wedge \Omega ,$$

this contradicts the fact that the scalar curvature s vanishes; integrating on M with the complex orientation and using the fact that c_1 can be thought of either as the the deRham class of $\frac{1}{2\pi}\mathrm{P}$ or as the Poincaré dual of $\frac{1}{m}[D]$, we have

$$\begin{aligned} 0 = \int_M s\Omega \wedge \Omega &= 4\int_M \mathrm{P} \wedge \Omega \\ &= 8\pi c_1 \cup [\Omega] \\ &= \frac{8\pi}{m}\int_D \Omega > 0 . \end{aligned}$$

The only way of avoiding this contradiction is to conclude that all the isolated fixed points have index +1. Condition (3) is therefore satisfied, and the action is docile. ∎

We will now justify our introduction of the of the notion of docility.

Definition 3 *An Einstein-Weyl manifold* $(X^3, [h], \mathbf{D})$ *is said to be* asymptotically hyperbolic *if there exists*

- *a connected Riemannian 3-manifold-with-boundary* $(Y, \tilde{h})$*;*
- *a non-degenerate defining function* $f : Y \to \mathbf{R}^+$ *of* ∂Y*; and*
- *a smooth 1-form* ν *on* Y *vanishing along* ∂Y*;*

such that $(X^3, [h], \mathbf{D})$ *is isomorphic to* $(Y - \partial Y)$ *equipped with the conformal class* $[\tilde{h}]$ *and the connection determined by the Levi-Civita connection* ∇ *of* $f^{-2}\tilde{h}$ *and the 1-form* ν *via formula (2).*

Lemma 1 *Suppose that* (M, g) *is a self-dual manifold with docile circle action. Then the interior* X *of the quotient* $Y = M/S^1$ *carries an asymptotically hyperbolic Einstein-Weyl structure which extends the one defined by equation (1) away from the fixed points.*

Proof. We begin by considering the geometry near a fixed surface. Using the exponential map relative to any conformal gauge, we can identify a neighborhood of a fixed surface Σ_j with its normal bundle. Letting (x^1, x^2) be local coordinates on Σ_j, letting r denote the distance from Σ_j, and letting t be an angular coordinate on the fibers of the normal bundle of Σ_j, the metric in a neighborhood of Σ_j is thus of the form

$$g = dr^2 + r^2(1+f)^2(dt + \theta_1 dx^1 + \theta_2 dx^2)^2 + u_{jk}dx^j dx^k ,$$

where f, θ_j, u_{jk}, $j, k = 1,2$ are smooth functions of (r, x^1, x^2) and are *even* as functions of r, and where, moreover, $f = 0$ along $r = 0$. The associated submersion metric is therefore

$$\hat{h} = dr^2 + u_{jk}dx^j dx^k ,$$

On the complement of Σ_j, let us now change conformal gauge by multiplying the metric by $1/r^2$, so that the submersion metric becomes the asymptotically hyperbolic metric

$$h = \frac{dr^2 + u_{jk}dx^j dx^k}{r^2}$$

and the Killing field $\xi = \frac{\partial}{\partial t}$ has length $1 + f$. Our formula (1) for ν then guarantees that, in this conformal gauge, is smooth up to $r = 0$ and vanishes there.

We now proceed by constructing the Hitchin mini-twistor space of the quotient Einstein-Weyl structure on a neighborhood of an isolated fixed point p at which the action is trivial on $\wedge_-$. Let us recall that the map induced on $S(\wedge_- M)$ by any isometry of a self-dual manifold (M, g) is a biholomorphic automorphism of the twistor space $Z = S(\wedge_-)$, and consequently any Killing field of M lifts to Z as the real part of a holomorphic vector field. At our isolated fixed-point p, the assumption on the induced action on $(\wedge_-)_p$ says that the holomorphic vector field ζ corresponding to the generator of the action vanishes along the twistor fiber L of p. Since $\Re\zeta$ generates the action of a compact group, the linear transformation $d\zeta : T_zZ \to T_zZ$ is diagonalizable at any $z \in L$; and since this transformation is an almost complex structure on the normal bundle determining the orientation opposite that of M, the eigenvalues of $d\zeta$ must be exactly $\{i, -i, 0\}$ at each point of L. Since $d\zeta$ is holomorphic along L, we obtain a holomorphic decomposition

$$TZ|_L \cong TL \oplus E_+ \oplus E_-$$

into the eigenspaces of $d\zeta$; here $E_\pm$ is the eigen-line-bundle of eigenvalue $\pm i$. But since the normal budle of a twistor line is $\mathcal{O}(1) \oplus \mathcal{O}(1)$, and the splitting type of a vector bundle on $\mathbf{CP}_1$ is unique, it follows that $E_+ \cong E_- \cong \mathcal{O}(1)$.

In the neighborhood of any point $z \in L$, let $S_\pm$ denote a hypersurface containing L, and whose tangent space along L is identically $TL \oplus E_\mp$. Let

$f_\pm$ be any nondegenerate defining function for $S_\pm$, so that $df_\pm$ is in the $\pm i$-eigen-bundle of $(d\zeta)^*$. We then set

$$u_\pm := \frac{1}{2\pi}\int_0^{2\pi} e^{\mp it}[\exp t\Re\zeta]^* f_\pm \, dt.$$

Let f_0 be any holomorphic function on a neighborhood of $z \in L$ which restricts to a local coordinate system on L centered at z, and set

$$u_0 := \frac{1}{2\pi}\int_0^{2\pi} [\exp t\Re\zeta]^* f_0 \, dt \ .$$

This produces a local coordinate system on Z near z in which S^1 acts by

$$(u_0, u_+, u_-) \mapsto (u_0, e^{it}u_+, e^{it}u_-) \ .$$

Now let $\mathcal{O}_{inv}$ denote the sheaf on L of germs of holomorphic functions on Z which are invariant under the S^1-action, or, equivalently, in the kernel of the holomorphic vector field ζ. Using the above coordinates, any such function is given near z by a convergent power series in (u_0, u_1), where $u_1 := u_+u_-$. It follows that $(L, \mathcal{O}_{inv})$ is the germ of an embedding of L in a complex surface $\mathcal{T}$. The inclusion $\mathcal{O}_{inv} \hookrightarrow \mathcal{O}_Z$ induces a projection from a neighborhood of $L \subset Z$ to a neighborhood of $L \subset \mathcal{T}$, explicitly given in the constructed coordinates by $(u_0, u_+, u_-) \mapsto (u_0, u_+u_-)$. The normal bundle of $L \subset \mathcal{T}$ is thus given by $E_+ \otimes E_- \cong \mathcal{O}(2)$. We also have a free anti-holomorphic involution $\sigma : \mathcal{T} \to \mathcal{T}$ induced by the real-structure of Z, since the latter commutes with the S^1-action. Letting Y denote the family of all σ-invariant rational curves in $\mathcal{T}$ of self-intersection 2, the Hitchin correspondence [20] [21] then equips Y with a natural Einstein-Weyl structure for which $\mathcal{T}$ is the space of oriented geodesics, and this will agree [22] with the submersion recipe away from the fixed point. ∎

Remark. In the bulk of this article, we have found it convenient to give the manifold-with-boundary $Y = M/S^1$ a smooth structure for which the quotient map $M \to Y$ becomes smooth. However, the use of r, rather than r^2, as a local coordinate on Y is incompatible with this! The present enlarging of the space of smooth functions of Y has the advantage that the conformal structure of $X = Y - \partial Y$ becomes smooth up to the boundary, and we shall adopt this convention in the sequel. This may be viewed as a technical, rather than fundamental, distinction, however; it is basically a matter of taste that we choose to use the Poincare' (conformal) model of hyperbolic space in dealing with points at infinity rather than using the Klein (projective) model, as would naturally occur if we continued to insist that the quotient map $S^4 \to B^3$ must be smooth! An important an useful technical aspect of the particular set of conventions we have just adopted, though, is that the operator $\star d\star(d-\nu)$ of an symptotically hyperbolic Einstein-Weyl structure now becomes totally characteristic at $r=0$ in the sense of Melrose.

We are now adequately prepared for the main results of this article.

Theorem 1 *Let M be a compact self-dual 4-manifold with (strictly) positive intersection form, and let g be a self-dual metric on M which admits a semi-free S^1-action. Then there exists a hyperbolic manifold X and a finite collection of points $\{q_1,\ldots,q_n\}\subset X$ from which (M,g) arises via the hyperbolic ansatz.*

Proof. We again set $Y=M/S^1$ and $X=Y-\partial Y$. By Lemma 1, X admits an Einstein-Weyl structure obtained by projection from M; moreover, this structure is asymptotically hyperbolic near ∂Y provided we remember to introduce a conformal factor on M which is singular at the fixed surface, as described above.

We will now solve a Dirichlet problem at ∞; namely, we seek a smooth function $\hat{V}:Y\to\mathbf{R}$ which solves the elliptic equation $d\star(d-\nu)\hat{V}=0$ on $X=Y-\partial Y$, with $\hat{V}\equiv 1$ on ∂Y. To do this, first notice that the function V arising from our action on M would be a solution of this problem were it not for the fact that it blows up at $\{q_1,\ldots,q_n\}$. If we cut off V by multiplying it by a smooth function which is $\equiv 0$ near $\{q_1,\ldots,q_n\}$ and $\equiv 1$ outside some small balls about these points, we therefore obtain a function v which solves the equation $d\star(d-\nu)v=\star f$, where f is smooth and compactly supported. It has been proven by Mazzeo [36] that $\Delta=\star d\star d$ is a Fredholm map $L^2_2\to L^2$ and our compact perturbation $\star d\star(d-\nu)$ of this operator is therefore Fredholm. But the L^2-adjoint of the later operator has no zeroth order term, and hence the kernel of the adjoint is trivial by the maximum priciple and the regularity theory at infinity of such totally characteristic operators [16]. Thus $d\star(d-\nu)w=-\star f$ for some $w\in L^2_2$, and we can now set $\hat{V}\equiv w+v$. By the same regularity theory, this is smooth and tends to 1 at infinity, as desired.

Now once we have our solution $\hat{V}$, let us observe that it must be positive. Indeed, on the set $X'=X-\{q_1,\ldots,q_n\}$, we may use V as a conformal factor for which the corresponding ν is co-closed, which eliminates the zeroth order term in the equation; thus $\hat{V}/V$ cannot have a maximum or a minimum on $X-\{q_1,\ldots,q_n\}$. The continuous function

$$\hat{V}/V:\overline{Y}\to\mathbf{R}$$

therefore has maximum 1 on ∂Y and minimum 0 at $\{q_1,\ldots,q_n\}$. (Recall that $V^{-1}=r+O(r^2)$ at p_j.) Thus $\hat{V}/V>0$, and hence $\hat{V}>0$ on X'. But we must also have $\hat{V}>0$ at $\{q_1,\ldots,q_n\}$; for otherwise, letting w denote some local positive solution near p_j, then, using w as a conformal factor, the corresponding ν is co-closed, so that $\hat{V}/w$ cannot have an interior minimum, as would certainly happen at p_j if we had $\hat{V}(p_j) = 0$. Hence $\hat{V} > 0$ on all of Y.

We now define a new half-conformally-flat 4-manifold $\hat{M}$, as follows: The potential $\hat{V}$ solves $d\star(d-\nu)\hat{V} = 0$ on all of (X, h), and extends smoothly to $Y = X \cup \Sigma$. But, by Proposition 1, the second Betti number of Y vanishes, so that the closed 2-form $\star(d-\nu)\hat{V}$ is the curvature of a connection 1-form θ on the trivial bundle $X \times S^1 \to X$. For a suitable conformal factor α, the metric

$$\hat{g} = \alpha(\hat{V}h + \hat{V}^{-1}\theta \otimes \theta)$$

now defines a self-dual metric on $X \times S^1$ which extends smoothly to the compact manifold $\hat{M} := (X \times S^1) \cup \partial Y$. But $\hat{M}$ now admits a semi-free circle action with fixed-point set ∂Y, and the normal bundle of this fixed point set is trivial. The signature of $\hat{M}$ is therefore zero, and the constructed metric $\hat{g}$ must be conformally flat.

Near a point p on the fixed surface ∂Y of the action, we can thus find a local conformal isometry between an open neighborhood $\mathcal{U}$ of p and an open subset $\mathcal{V}$ of S^4. The conformal Killing field ξ on $\mathcal{U}$ then becomes a conformal Killing field on $\mathcal{V}$, and by Liouville's theorem [34], therefore extends as an infinitesimal Möbius transformation of S^4. Since ξ is periodic, so is the corresponding vector field on S^4, and we have a homomorphism $S^1 \to SO(5,1)$, which must take its image in a subgroup conjugate to the maximal compact $SO(5)$. Our Killing field ξ thus corresponds a periodic Killing field of the standard metric on S^4, and, like ξ, this Killing field vanishes along a (totally geodesic) surface. The infinitcsimal Möbius transformation in question is thus simply a rotation

$$\begin{bmatrix} \cos t & \sin t & 0 & 0 & 0 \\ -\sin t & \cos t & 0 & 0 & 0 \\ 0 & 0 & 1 & 0 & 0 \\ 0 & 0 & 0 & 1 & 0 \\ 0 & 0 & 0 & 0 & 1 \end{bmatrix}$$

of S^4 around a great 2-sphere. The quotient Einstein-Weyl structure $(X, h, \mathbf{D})$ is therefore hyperbolic near the surface at infinity $\partial Y \subset Y$. In particular, the connection on $\wedge^3 TX$ induced by $\mathbf{D}$ is flat near ∂Y. But since any Einstein-Weyl structure is real-analytic in a suitable system of charts, it follows that the connection on $\wedge^3 TX$ induced by $\mathbf{D}$ is everywhere flat. Since X is orientable, this gives us a global volume form invariant under $\mathbf{D}$, which is therefore the Levi-Civita connection of the unique metric $\hat{h}$ in the conformal class $[h]$ with this volume element. Since this metric has constant negative curvature near infinity, real-analyticity guarantees that its sectional curvature is a negative constant, which may be taken to be -1 by multiplying $\hat{h}$ by a positive constant as necessary.

To finish, we need merely observe that the potential V arising from the submersion construction satisfies $\Delta V = 0$ on the complement of the finite set $\{q_1, \ldots, q_n\}$, with $V \sim \frac{1}{2r_j}$ near q_j, whereas $V \to 1$ at ∂Y. It follows that $V - 1$ is the sum of the Green's functions of the points $q_1, \ldots, q_n$. ∎

Theorem 2 *Let (M, g) be a compact half-conformally-flat Riemannian manifold which admits a semi-free conformal circle action. Suppose that (M, g) is not conformally flat and has non-negative scalar curvature. Then there exists a hyperbolic manifold X and a finite collection of points $\{q_1, \ldots, q_n\} \subset X$ from which (M, g) arises via the hyperbolic ansatz. Moreover, X is either hyperbolic 3-space, a handle-body, or the determinant bundle over a surface of negative Euler characteristic.*

Proof. We must first prove that X is hyperbolic. For this, the only case not covered by Theorem 1 is that of a compact scalar-flat Kähler surface. We now continue the analysis of this case which we began in Proposition 2. As we already saw in that discussion, the quotient M/S^1 can be canonically identified with $[-1, 1] \times \Sigma_{\mathbf{g}}$ for some compact complex curve $\Sigma_{\mathbf{g}}$ in such a way that projection $M \to \Sigma_{\mathbf{g}}$ is just the representation of M as a ruled surface, and the projection $M \to [-1, 1]$ given by $Af + B$ for some real constants A and B. Indeed, by multiplying the metric by a constant and adding a constant to f, we may assume that f is precisely the projection $M \to [-1, 1]$. Note that the preceding renormalization of the metric is precisely that for which the area of a fiber of $M \to \Sigma_{\mathbf{g}}$ has area 4π.

If we now equip $Y = M/S^1 = [-1, 1] \times \Sigma_{\mathbf{g}}$ with local coordinates (x, y, z), where $z = f$ and $x + iy$ is any holomorphic coordinate on $\Sigma_{\mathbf{g}}$, we can analyze our scalar-flat Kähler metric on M in terms of a specialized form of the Einstein-Weyl machinery, developed in [29]. Namely, simply by virtue of the fact that we have a Kähler metric with a circle action whose Hamiltonian is z, we can write it in the form

$$g = e^u v(dx^2 + dy^2) + v dz^2 + v^{-1}\theta^2$$

where θ is the connection of a circle bundle over Y minus a finite number of points, with the curvature of θ given by

$$d\theta = v_x dy \wedge dz + v_y dz \wedge dx + (ve^u)_z dx \wedge dy \; .$$

The fact that this is closed then yields

$$v_{xx} + v_{yy} + (ve^u)_{zz} = 0. \tag{7}$$

On the other hand, the fact that our metric is *scalar-flat* becomes the equation

$$u_{xx} + u_{yy} + (e^u)_{zz} = 0 \; . \tag{8}$$

The so-called *Toda-lattice equation* (8) is actually just a simplified version of the Einstein-Weyl equations. Namely, given a function u on a region X of

$\mathbf{R}^{\not\#}$ satisfying (8), we may endow X with a conformal structure determined by the Riemannian metric

$$\hat{h} := e^u(dx^2 + dy^2) + dz^2,$$

and with the torsion-free connection $\mathbf{D}$ defined by

$$\mathbf{D}_\xi \eta := \nabla_\xi \eta + \nu(\xi)\eta + \nu(\eta)\xi - \hat{h}(\xi, \eta)\nu^\#,$$

where $\nu := -u_z dz$, $\nu^\# = -u_z \frac{\partial}{\partial z}$, and ∇ denotes the Levi-Civita connection of $\hat{h}$. The function v, on the other hand, corresponds to V in the general submersion construction. In particular, the 2-form

$$v_x dy \wedge dz + v_y dz \wedge dx + (ve^u)_z dx \wedge dy$$

is globally defined on $\Sigma_{\mathbf{g}} \times [-1, 1]$, and the integrality condition becomes the requirement that

$$\frac{1}{2\pi} \int_{\Sigma_{\mathbf{g}} \times \{c\}} (ve^u)_z dx \wedge dy \in \mathbf{Z}$$

for any constant $c \in [-1, 1]$ for which $\Sigma_{\mathbf{g}} \times c$ is in the domain of v. Let us rewrite this condition as

$$\frac{1}{2\pi} \frac{d}{dz} \int_{\Sigma_{\mathbf{g}} \times \{z\}} ve^u dx \wedge dy \in \mathbf{Z}.$$

Now note that the differential 2-form $e^u dx \wedge dy$ is globally defined on $\Sigma_{\mathbf{g}} \times [-1, 1]$, and we may think of it as a z-dependent Kähler form on $\Sigma_{\mathbf{g}}$. Equation (8)then precisely says that the second derivative of this metric (with respect to z) is given by its Ricci form. If we integrate (8) over $\Sigma_{\mathbf{g}}$ and use the Gauss-Bonnet theorem, we thus conclude that

$$\frac{d^2}{dz^2} \int_{\Sigma_{\mathbf{g}} \times \{z\}} e^u dx \wedge dy = 8\pi(1 - \mathbf{g}),$$

and, in light of the fact that $e^u \to 0$ at $z \pm 1$ (because $v^{-1} \to 0$, $ve^u \not\to 0$ at the fixed-point set), we must have

$$\int_{\Sigma_{\mathbf{g}} \times \{z\}} e^u dx \wedge dy = 4\pi(\mathbf{g} - 1)(1 - z^2).$$

We now introduce a new solution $\hat{v}$ of equation (7) by requiring that $\hat{v}$ be defined on all of $X := (\Sigma_{\mathbf{g}} \times]-1, 1[)$ and that $\hat{v} \sim \frac{1}{2(1\mp z)}$ near $z = \pm 1$. We can do this by the same argument used to produce $\hat{V}$ in the proof of Theorem 1; indeed, by setting $\hat{v} = (1-z^2)^{-1}\hat{V}$, the conformal change $\hat{h} \mapsto h := (1-z^2)^{-2}\hat{h}$ precisely reduces this to the previous problem. We then have

$$\begin{aligned} \lim_{z \to \pm 1} \int_{\Sigma_{\mathbf{g}} \times \{z\}} \hat{v} e^u dx \wedge dy &= \lim_{z \to \pm 1} [4\pi(\mathbf{g} - 1)(1 - z^2)][\frac{1}{2(1 \mp z)}] \\ &= 4\pi(\mathbf{g} - 1). \end{aligned}$$

On the other hand, equation (7) implies that

$$\frac{d}{dz}\int_{\Sigma_{\mathbf{g}}\times\{z\}} \hat{v}e^u dx\wedge dy = k$$

for some constant k. Since

$$\lim_{z\to 1}\int_{\Sigma_{\mathbf{g}}\times\{z\}} \hat{v}e^u dx\wedge dy = \lim_{z\to -1}\int_{\Sigma_{\mathbf{g}}\times\{z\}} \hat{v}e^u dx\wedge dy,$$

it follows that $k=0$, and $(\hat{v}_x dy\wedge dz+\hat{v}_y dz\wedge dx+(\hat{v}e^u)_z dx\wedge dy)$ represents the trivial deRham class on X. Hence this 2-form is the curvature of a connection θ on the trivial bundle $X\times S^1\to X$. The metric

$$\hat{g} = (\hat{v}h+\hat{v}^{-1}\theta\otimes\theta)$$

now defines a self-dual metric on $X\times S^1$ which extends smoothly to the compact manifold $\hat{M} := (X\times S^1)\cup\partial Y$. Just as in Theorem 1, $\hat{M}$ now admits a semi-free circle action with fixed-point set ∂Y, and the normal bundle of this fixed point set is trivial. The signature of $\hat{M}$ is therefore zero, and the constructed metric $\hat{g}$ is conformally flat. Proceeding exactly as in the previous proof, we then conclude that X is hyperbolic, and the potential $\hat{V}$ is of the form $1+\sum G_j$.

To finish, we need to classify the possible hyperbolic manifolds $X=\mathcal{H}^3/\Gamma$ that might occur. Generalizing Braam's analysis [5] of the conformally flat case, Jongsu Kim [23] has shown that the Hausdorff dimension of the limit set Λ of Γ may be estimated in terms of the scalar curvature s of any half-conformally-flat 4-manifold M arising from the hyperbolic ansatz on X; specifically, $\dim_{\mathcal{H}}\Lambda\leq 1$ if $s\geq 0$. Using Maskit's classification theorem's [35], this implies that Γ is either trivial, Schottky, Fuchsian, or extended Fuchsian. These possibilities correspond to the listed possibilities for X. ∎

Combining the above with Propositions 1 and 2, we immediately deduce the following corollaries:

Corollary 1 *A self-dual metric on $n\mathbf{CP}_2$ has Moishezon twistor space if it admits a semi-free isometric circle-action.*

Corollary 2 *Let (M,g) be a compact self-dual manifold with non-negative Yamabe constant and which admits a semi-free isometric circle action. Assume (M,g) is not conformally flat. Then M is diffeomorphic either to a connected sum $k(S^1\times S^3)\#n\mathbf{CP}_2$, $n>0$, $k\geq 0$, or to a connected sum $B\#n\mathbf{CP}_2$, where B is an S^2-bundle over a surface of negative Euler characteristic.*

Remarks.

- The hypothesis that the action be semi-free seems to be essential for any result along the lines of Theorem 1. In particular, by considering Example 2 in the case when the n points are all collinear, one can construct self-dual metrics on $n\mathbf{CP}_2$ with an isometric $S^1 \times S^1$ action such that the action of the second factor is *not* semi-free and such that the corresponding Einstein-Weyl geometry is *not* hyperbolic. It would now be interesting to try to deform these examples in such a manner as to destroy the semi-free circle-action while at the same time preserving the "bad" action, thereby showing that the semi-free hypothesis of the above theorem is essential.

- In a similar vein, examples like that of rotation of S^4 around two isolated fixed points strongly indicate our hypothesis regarding the action on Λ_- at isolated fixed points is again essential.

- Is every asymptotically hyperbolic Einstein-weyl manifold actually hyperbolic? One might take Theorems 1 and 2 as evidence in favor of such a theorem. Indeed, there is a reasonable hope that a beautiful series of arguments of Tod [43], who recently classified *compact* Einstein-Weyl 3-manifolds, might be generalized in order to prove this. Such a proof would involve showing, by a Bochner-type argument, that $\nu^{\#}$ must be a Killing field of the metric $\hat{V}^2 h$, and appealing to Tod's local classification of Einstein-Weyl spaces for which $\nu^{\#}$ is Killing. Such an approach would eliminate the need to show that the cohomology class $[\frac{1}{2\pi} \star (d-\nu)\hat{V}]$ is integral, which is the key step of the above arguments.

5 Application: Critical Kähler Metrics

Given a compact complex manifold (M, J) and a Kähler class $[\Omega] \in H^{1,1}(M)$, one may ask whether this class contains a metric of constant scalar curvature. This problem was first posed by Calabi, who went on to prove any number of substantial results on the subject (cf. e.g. [6]), including the fact that the set of solutions, modulo automorphisms of (M, J), is *discrete.* In this section, we will prove *existence and uniqueness* results for this problem in the case when (M, J) is a compact complex surface with a non-trivial holomorphic vector field and when the Kähler class $[\Omega]$ satisfies $c_1 \cdot [\Omega] = 0$. Our results will follow from Theorem 1 in conjunction with the results of [29] and [31].

The relevance to this problem of the previous discussion stems from the observation [13] that a Kähler manifold of complex dimension 2 is *anti-self-dual* with respect to the standard orientation iff its scalar curvature is identically zero. In twistor terms, this class of half-conformally-flat manifolds is characterized [38] by the existence of an effective divisor D in the twistor space Z such that $-2([D]+[\overline{D}]) = \kappa$ and $[D]\cdot[\overline{D}] = 0$, where $\overline{D}$ denotes the image of D under the real structure σ.

Of course most complex surfaces cannot possibly admit such a metric, since one would then have $0 = \int_M s\Omega \wedge \Omega = c_1 \cdot [\Omega]$, and this implies (cf. [45] and/or the end of the proof of Proposition 2) and that either $H^0(M, \kappa^m) = 0$ $\forall m \neq 0$ or else $c_1^{\mathbf{R}}(M) = 0$, and in the former case the Kodaira-Enriques classification implies that M is rational or ruled. Another potential obstruction stems from the observation of Matsushima/Lichnerowicz [33] that the connected component of the group of biholomorphisms of a compact Kähler manifold of constant scalar curvature must have a compact real form. If we further require that this group be non-trivial, there must in particular be a holomorphic action of $\mathbf{C}^*$ on M. But this already narrows things down enormously:

Lemma 2 *Let M be a compact complex surface admitting a holomorphic $\mathbf{C}^*$-action and a Kähler class $[\Omega]$ such that $c_1 \cdot [\Omega] = 0$. Suppose, moreover, that M is not finitely covered by a a complex torus. Then, for some holomorphic line bundle $\mathcal{L} \to \Sigma_{\mathbf{g}}$ over a compact complex curve $\Sigma_{\mathbf{g}}$ of genus ≥ 2, M is obtained from the minimal ruled surface $\mathsf{P}(\mathcal{L} \oplus \mathcal{O}) \to \Sigma_{\mathbf{g}}$ by blowing up points along the zero section of $\mathcal{L} \subset \mathsf{P}(\mathcal{L} \oplus \mathcal{O})$. Moreover, the given action is precisely that generated by the Euler vector field on $\mathcal{L}$.*

Proof. By the above-mentioned plurigenera-vanishing-theorem of Yau [45], the fact that $c_1 \cdot [\Omega] = 0$ implies that either $c_1^{\mathbf{R}}(M) = 0$ or else M is either rational or ruled. Since $h^0(\mathbf{K3}, \mathcal{O}(T\mathbf{K3})) = h^{1,0}(\mathbf{K3}) = \frac{1}{2}b_1(\mathbf{K3}) = 0$, M cannot be covered by a K3-surface; the Kodaira-Enriques classification then says that either $c_1^{\mathbf{R}}(M) \neq 0$ or M is finitely covered by a complex torus[2]. Since we have explicitly required that M not be finitely covered by a complex torus, it follows that $c_1^{\mathbf{R}}(M) \neq 0$. We may thus conclude that M is either $\mathbf{CP}_2$ or a (possibly non-minimal) ruled surface $M \to \Sigma_{\mathbf{g}}$ with typical fiber $\mathbf{CP}_1$, where $\Sigma_{\mathbf{g}}$ is some compact complex curve.

Since $c_1 \cdot [\Omega] = 0$, $c_1^{\mathbf{R}}$ is a non-zero primitive class in $H^{1,1}$, and $c_1^2 < 0$. Thus, with respect to the complex orientation, $2\chi + 3\tau < 0$, in particular excluding $\mathbf{CP}_2$. If the curve $\Sigma_{\mathbf{g}}$ has genus < 2, we simultaneously obtain an estimate on the number of exceptional divisors contained by M. Specifically, if M is obtained by blowing up a minimal rational ruled surface $F_k \to \mathbf{CP}_1$ at m points, we must have $k \geq 9$; and if M is instead obtained by blowing

[2]Recall that the compact complex surfaces which are non-trivially covered by a complex torus are precisely the *hyper-elliptic surfaces*, all of which are finitely covered by a product of two elliptic curves; these fall into seven classes, depending on the action of the covering group. Each such surface *does* actually admit a non-trivial holomorphic vector field, corresponding to the the vertical tangent bundle of a realization of the surface as an elliptic fibration over $\mathbf{CP}_1$.

up a minimal ruled surface $\check{M} \to \mathsf{E}$ over an elliptic curve $\mathsf{E} = \mathbf{C}/\Lambda$ at m points, then $m > 0$.

Let ξ denote a holomorphic vector field on M whose real part generates a S^1-action, and let $\pi : \check{M} \to \Sigma_{\mathbf{g}}$ denote a $\mathbf{CP}_1$-bundle from which M can be obtained by a blow-up $\wp : M \to \check{M}$. We then consider the component of $\wp_*\xi$ normal to the fibers of π. Since the normal bundle of each such fiber is trivial, this normal component is constant up the fibers, so that $(\pi\wp)_*\xi$ is a well-defined holomorphic vector field on $\Sigma_{\mathbf{g}}$. If $\Sigma_{\mathbf{g}}$ has genus > 1, this vector field must vanish. If, on the other hand, $\Sigma_{\mathbf{g}}$ has genus 1, the fact that $\wp$ involves blowing up at at least one point forces $\wp_*\xi$, and hence $(\pi\wp)_*\xi$, to have at least one zero, implying that $(\pi\wp)_*\xi \equiv 0$. Finally, if $\Sigma_{\mathbf{g}}$ has genus 0, and if $(\pi\wp)_*\xi \not\equiv 0$, the $m \geq 9$ blown-up points of $\pi : F_k \to \mathbf{CP}_1$ must be located on at most 2 fibers of π; but then $F_k = \mathcal{O} \oplus \mathcal{O}(\urcorner)$ admits sections of κ^{-m} which vanishes along these two fibers to order m, and this section lifts to a non-zero element of $H^0(M, \mathcal{O}(\kappa^{-m}))$, contradicting our observation that this group must vanish when $c_1 \cdot [\Omega] = 0$. The vector field $(\pi\wp)_*\xi$ must therefore vanish identically, and $\wp_*\zeta$ is tangent to the fibers of π.

Now the model $\pi : \check{M} \to \Sigma_{\mathbf{g}}$ may be represented in the form $\mathbf{P}(E) \to \Sigma_{\mathbf{g}}$ for a rank 2 holomorphic vector bundle $E \to \Sigma_{\mathbf{g}}$ which is completely specified once an arbitrary line bundle $\wedge^2 E$ is chosen, subject to the condition $c_1(E) \equiv w_2(\pi)$ (mod2). The vector field $\wp_*\zeta$ is then uniquely specified by a trace-free holomorphic section A of $\mathcal{E}nd(E)$. The determinant of A is then a holomorphic function on $\Sigma_{\mathbf{g}}$, hence a constant. On the other hand, since the real part of ζ is periodic, A is diagonalizable, and A must be a half-integer multiple of

$$\begin{bmatrix} -i & 0 \\ 0 & i \end{bmatrix}.$$

E thus globally splits as a direct sum of the eigenspaces of A, and, twisting by a line bundle, we may therefore take $E = \mathcal{L} \oplus \mathcal{O}$, so that ζ becomes a constant multiple of the Euler vector field on $\mathcal{L}$. The blown-up points must all occur at zeroes of ζ, namely either at the zero section of $\mathcal{L}$ or a the "infinity section" corresponding to the $\mathcal{O}$ factor. The latter possibility may be reduced to the former by noticing that the proper transform of a fiber through exactly one blown-up point is a (-1)-curve, which may therefore be blown down, thereby leading to a different minimal model. In our case, iteration of this procedure allows us to replace blown-up points "at the infinity section" by a blown-up points "at the zero section," at the small price of twisting our line bundle $\mathcal{L}$ by the divisor of the relevant points of $\Sigma_{\mathbf{g}}$. ∎

Now suppose that (M, g, J) is a compact scalar-flat Kähler surface which admits a non-trivial holomorphic vector field and is not covered by a torus. By Lichnerowicz [33], the connected component of the isometry group of M is a compact real form for a central extension of the connected component of the complex automorphism group, and M therefore admits an isometric circle action generated by the real part of a holomorphic vector field. By Lemma 2, this action is semi-free. Lichnerowicz's proof, moreover, says that this circle action, thought of as a group of automorphisms of the symplectic manifold (M, Ω), is generated by a Hamiltonian f, called the holomorphy potential of the vector field. Indeed, as we already saw in the proof of Theorem 2, after emultiplying the metric by a constant so as to make the area of a fiber of $M \to \Sigma_{\mathbf{g}}$ equal 4π, the quotient M/S^1 can be canonically identified with $[-1,1] \times \Sigma_{\mathbf{g}}$ for some compact complex curve $\Sigma_{\mathbf{g}}$ in such a way that projection $M \to \Sigma_{\mathbf{g}}$ is just the representation of M as a ruled surface, and the projection $M \to [-1,1]$ is a Hamiltonian for the the Killing field. If we now equip $Y = M/S^1 = [-1,1] \times \Sigma_{\mathbf{g}}$ with local coordinates (x, y, z), where $z = f$ and $x + iy$ is any holomorphic coordinate on $\Sigma_{\mathbf{g}}$, our scalar-flat Kähler metric on M is completely specified by a solution of the Toda equation

$$u_{xx} + u_{yy} + (e^u)_{zz} = 0,$$

which specifies the Einstein-Weyl structure, together with a solution of the linear equation

$$v_{xx} + v_{yy} + (ve^u)_{zz} = 0.$$

Now we proved in Theorem 2 that our Einstein-Weyl structure must actually be a hyperbolic structure in a peculiar gauge, so that the defining 1-form $\nu := -u_z dz$ of the structure must be exact. In particular, $d\nu = 0$, so that $u_{zx} = u_{zy} = 0$, and $u = a(z) + b(x, y)$, where a and b are uniquely determined once we impose the additional requirement that $a(0) = 0$. The Toda equation (8) now simplifies to become

$$b_{xx} + b_{yy} + e^b (e^a)_{zz} = 0 \ .$$

Hence, for some constant k we have

$$b_{xx} + b_{yy} + ke^b = 0 \tag{9}$$

$$(e^a)_{zz} = k \ . \tag{10}$$

The equation 9 states that $e^b(dx^2 + dy^2)$ has constant sectional curvature $\frac{k}{2}$, whereas equation 10 yields $a = \log(\frac{k}{2}z^2 + k_1 z + k_2)$. We can now eliminate ν by changing Weyl gauge, replacing $\hat{h}$ with $h = (\frac{k}{2}z^2 + k_1 z + k_2)^{-1}\hat{h}$, which must therefore be a metric of constant negative curvature on $[-1,1] \times \Sigma_{\mathbf{g}}$, with $z = \pm 1$ corresponding to infinity for this manifold. We therefore have $\frac{k}{2}z^2 + k_1 z + k_2 = \frac{k}{2}(z^2 - 1)$, where $k < 0$. Since our conventions set $a(0) = 0$, it follows that $k = -2$, and $a = \log(1 - z^2)$. It of course also also follows that that $e^b(dx^2 + dy^2)$ has constant sectional curvature -1. Since the metric

$$\hat{h} := e^u(dx^2 + dy^2) + dz^2$$

can be constructed in a manner independent of the choice of (x, y), namely, as the unique metric in the submersion conformal class with z-component

1, and since $b = u - \log(z^2 - 1)$, it follows that $e^b(dx^2 + dy^2)$ is globally defined on Σ_g, and is therefore *a fortiori* the unique Hermitian metric on Σ_g of curvature -1. Let us call this canonical metric g_Σ. The hyperbolic metric h on $Y - \partial Y$ arising from the submersion construction is now just

$$h = \frac{\hat{h}}{z^2 - 1} = \frac{g_\Sigma}{1 - z^2} + \frac{dz^2}{(1 - z^2)^2} \,. \tag{11}$$

The upshot of all this is that any compact scalar-flat Kähler surface arises from the construction first explained in [31]. Given a finite collection $\{q_1, \dots, q_m\}$ of points in $[-1, 1] \times \Sigma_g$, let G_j be the associated Green's for the hyperbolic metric (11), and set $V = 1 + \sum_{j=1}^m G_j$. Notice that the fact that $\Delta V = 0$ away from $\{q_1, \dots, q_m\}$ amounts to the fact that $v = (1 - z^2)V$ solves (7); thus the hyperbolic ansatz in this case automatically produces scalar-flat Kähler surfaces. In general, however, the 2-form $\frac{1}{2\pi} \star dV$ will fail to be in an integral deRham class. In fact, a necessary and sufficient condition for this to be true is that $\frac{1}{2}\sum_{j=1}(1 + z_j) \in \mathbb{Z}$, where z_j is the z-coordinate of q_j. If this is satisfied, there will be a continuum of circle bundles with connection θ over $([-1, 1] \times C) - \{q_1, \dots, q_m\}$ with curvature $\star dV$, and for each of these we can now carry out the hyperbolic ansatz, equipping the resulting half-conformally flat manifold with the complex structure J determined by $dx \mapsto dy$ and $dz \mapsto \frac{1-z^2}{V}\theta$. For any positive constant K, the metric $g = K(1 - z^2)(Vh + V^{-1}\theta^2)$ now becomes a scalar-flat Kähler metric with respect to this complex structure. The vertical line segments joining the q_j to the surface $\{-1\} \times \Sigma_g$ have as their inverse images exceptional curves which may be blown down to produce a minimal model of our surface; the value of the Kähler class on each such curve is just the length of the corresponding line segment times $2K\pi$, whereas the area of any fiber of $M \to \Sigma_g$ is, by the same "principal of Archimedes," exactly $4K\pi$. Our minimal model is just $\mathsf{P}(\mathcal{L} \oplus \mathcal{O})$ for some holomorphic line bundle of degree $\frac{1}{2}\sum_{j=1}(1 + z_j)$ on Σ_g, and the blown-up points all occur at the zero section of $\mathcal{L}$. The space of flat circle bundles on Σ_g acts transitively by tensor product on the circle bundles on $([-1, 1] \times C) - \{q_1, \dots, q_m\}$ with curvature $\star dV$, and it is easy to see that this exactly corresponds to varying $\mathcal{L}$ through all holomorphic line bundles on Σ_g with degree $\frac{1}{2}\sum_{j=1}(1 + z_j)$. Putting this all together, we deduce the following result:

Theorem 3 *Let $\mathcal{L} \to \Sigma_g$ be a holomorphic line bundle over a compact complex curve Σ_g, and let (M, J) be a ruled surface obtained from the minimal ruled surface $\mathsf{P}(\mathcal{L} \oplus \mathcal{O}) \to \Sigma_g$ by blowing up m distinct points along the zero section of $\mathcal{L} \subset \mathsf{P}(\mathcal{L} \oplus \mathcal{O})$. Let $[\Omega]$ be any Kähler class on M such that $c_1 \cdot [\Omega] = 0$. Let $\{E_j\}$ denote the exceptional divisors corresponding to the blown-up points and let F denote the the fiber class of the projection $M \to \Sigma_g$. Then there is a Kähler metric of constant scalar curvature in the class $[\Omega]$ iff*

$$\sum_j \int_{E_j} [\Omega] = (deg\ \mathcal{L}) \int_F [\Omega] \ . \tag{12}$$

When such a metric exists, it is unique modulo automorphisms of (M, g).

Remarks.

1. Such Kähler classes $[\Omega] \in H^{1,1}(M)$ exist iff $\Sigma_{\mathbf{g}}$ has genus $\mathbf{g} \geq 2$ and $m \neq 1$.

2. Since $c_1 \cdot [\Omega] = 0$, the above condition on $[\Omega]$ is equivalent to the condition that
$$\int_C \Omega = (\mathbf{g} - 1) \int_F \Omega \ ,$$
where $\mathbf{g}$ is the genus of $\Sigma_{\mathbf{g}}$.

3. If some of the blown-up points coincide, the result still holds provided one counts the exceptional curves resulting from the iterated blow-ups with suitable multiplicities.

Corollary 3 *Let M be a compact complex surface with a non-trivial holomorphic vector field. Then M admits a scalar-flat Kähler metric iff M is one of the following:*

1. *a complex torus $\mathbb{C}^2/\Lambda$; or*

2. *a minimal hyperelliptic surface; or*

3. *a minimal ruled surface $\mathbf{P}(\mathcal{L} \oplus \mathcal{O}) \to \Sigma_{\mathbf{g}}$, where $\Sigma_{\mathbf{g}}$ has genus $\mathbf{g} \geq 2$ and $\mathcal{L} \to \Sigma_{\mathbf{g}}$ has degree 0; or*

4. *a non-minimal ruled surface obtained from the minimal model $\mathbf{P}(\mathcal{L} \oplus \mathcal{O}) \to \Sigma_{\mathbf{g}}$, where genus $\mathbf{g} \geq 2$, deg $\mathcal{L} > 0$, by blowing up $m >$ deg $\mathcal{L}$ points along the zero section of $\mathcal{L}$.*

Conjecture. The obstruction (12), i.e. $[\Omega]\cdot(-\deg \mathcal{L})[F] + \sum [E_j])$, occurring in Theorem 3 is precisely the Futaki invariant of $(M, J, [\Omega], \zeta)$.

Acknowledgements. The author would like to thank Gang Tian, Paul Tod, Nigel Hitchin, Max Pontecorvo, Yat-Sun Poon, Henrik Pedersen, Santiago Simanca and Jongsu Kim for a variety of helpful conversations, and the Taniguchi Foundation for providing an opportunity to visit to Japan in such delightful mathematical company.

References

[1] M. Anderson, P. Kronheimer, and C. LeBrun, "Ricci-Flat Kähler Manifolds of Infinite Topological Type," **Comm. Math. Phys. 125** (1989) 637-642

[2] M. Atiyah, N.J. Hitchin and I. M. Singer,"Self-Duality in Four Dimensional Riemannian Geometry," **Proc. R. Soc. Lond. A 362** (1978) 425-461.

[3] M. Atiyah and I. M. Singer,"Index of Elliptic Operators: III," **Proc. R. Soc. Lond. A 362** (1978) 425-461.

[4] R. Bott, "Non-Degenerate Critical Manifolds," **Ann. Math. 60** (1954) 248-261.

[5] P. J. Braam, "A Kaluza-Klein Approach to Hyperbolic Three-Manifolds,"**L'Enseignement Mathématique 34** (1988) 275-311

[6] E. Calabi, "Extremal Kähler Metrics," in **Seminar on Differential Geometry** Ann. Math. Studies 102 (S.T. Yau, Editor), Princeton University Press, Princeton, NJ, 1982, pp. 259-290.

[7] E. Cartan, "Sur une Classe d'Espaces de Weyl," **Ann. Scient. Ec. Norm. Sup. 60** (1943) 1-16.

[8] F. Campana, "On Twistor Spaces of the Class $\mathcal{C}$," **J. Differential Geometry**, *to appear.*

[9] F. Campana, C. LeBrun, and Y.-S. Poon, "On Complex Manifolds Bimeromorphic to Kähler Manifolds,"*preprint*

[10] S. K. Donaldson and R. Friedman, "Connected Sums of Self-Dual Manifolds and Deformations of Singular Spaces," **Nonlinearity 2,** (1989) 197-239.

[11] R. Fintushel, "Circle Actions on Simply Connected 4-Manifolds," **Trans. Am. Math. Soc. 230** (1977) 147-171.

[12] A. Floer, "Self-Dual Conformal Structures on $\ell\mathbb{CP}^2$," **J. Diff. Geometry**, *to appear.*

[13] P. Gauduchon, "Surfaces Kählériennes dont la Courbure Vérifie Certaines Conditions de Positivité," in **Géometrie Riemannienne en Dimension 4**. *Séminaire A. Besse, 1978/1979*, (Bérard-Bergery, Berger, and Houzel, eds.), CEDIC/Fernand Nathan, 1981.

[14] G. W. Gibbons and S. W. Hawking, "Gravitational Multi-Instantons," **Phys. Lett. 78B** (1978) 430-432.

[15] G. W. Gibbons and S. W. Hawking, "Classification of Gravitational Instanton Symmetries," **Comm. Math. Phys. 66** (1979) 291-310.

[16] C. R. Graham, "The Dirichlet Problem for the Bergman Laplacian II," **Comm. PDE 8** (1983) 563-641.

[17] N. Hitchin, "On the Curvature of Rational Surfaces," in **Proceedings of Symposia in Pure Mathematics 27**, American Mathematical Society, Providence, RI, 1975, pp 65-80.

[18] N. J. Hitchin, "Polygons and Gravitons," **Math. Proc. Cambridge Philos. Soc. 83** (1979) 465-476.

[19] N. J. Hitchin, "Kählerian Twistor Spaces," **Proc. Lond. Math. Soc. 43** (1981) 133-150.

[20] N. J. Hitchin, "Complex Manifolds and Einstein's Equations," *Lecture Notes in Mathematics* **970** (1982) 73-99.

[21] N. J. Hitchin,"Monopoles and Geodesics," **Comm. Math. Phys. 83** (1982) 579-602.

[22] P.E. Jones and K.P. Tod, "Minitwistor Spaces and Einstein-Weyl Spaces," **Class. Quantum Grav. 2** (1985) 565-577.

[23] J.S. Kim, "Generalized Construction of Self-Dual Metrics by Hyperbolic Manifolds," *preprint*, 1990.

[24] P. Kronheimer,"The Construction of ALE Spaces as Hyper-Kähler Quotients," **J. Diff. Geometry 29** (1989) 665-683.

[25] H.N. Kuiper,"On Conformally Flat Spaces in the Large," **Ann. Math. 50**(1949) 916-924.

[26] J. Lafontaine, "Remarques sur les Variétés Conformément Plates," **Math. Ann. 259** (1982)313-319.

[27] C. LeBrun, "$\mathcal{H}$-Space with a Cosmological Constant," **Proc. R. Soc. London A 380**(1982)171-185

[28] C. LeBrun, "On the Topology of Self-Dual Manifolds," **Proc. Am. Math. Soc. 98** (1986) 637-640.

[29] C. LeBrun, "Explicit Self-Dual metrics on $\mathbb{CP}_2\#\cdots\#\mathbb{CP}_2$," **J. Diff.Geometry**, *to appear.*

[30] C. LeBrun, "Anti-Self-Dual Hermitian Metrics on Blown-Up Hopf Surfaces,"**Math. Ann.**, *to appear.*

[31] C. LeBrun, "Scalar-Flat Kähler Metrics on Blown-Up Ruled Surfaces," **J. reine u. angew. Math.**, *to appear.*

[32] C. LeBrun, "Asymptotically Flat Scalar-Flat Kähler Surfaces," *preprint* 1990.

[33] A. Lichnerowicz, "Sur les Transformations Analytiques des Variétés Kähleriennes," **C. R. Acad. Sci. Paris 244** (1957) 3011-3014.

[34] S. Lie, "Inbesondere Linien und Kugelkomplexe mit Anwendung auf die Theorie partieller Differentialgleichungen," **Math. Ann. 5** (1872) 145-246.

[35] B. Maskit, **Kleinian Groups**, Springer-Verlag, 1988.

[36] R. R. Mazzeo, "The Hodge Cohomology of a Conformally Compact Metric," **J. Diff. Geo. 28** (1988) 309-339.

[37] R. Penrose, "Non-linear Gravitons and Curved Twistor Theory," **Gen. Rel. Grav. 7** (1976) 31-52.

[38] M. Pontecorvo, "On Twistor Spaces of Anti-Self-Dual Hermitian Surfaces," **Trans. Am. Math. Soc.** *to appear.*

[39] Y.S. Poon, "Compact Self-Dual Manifolds with Positive Scalar Curvature," **J. Differential Geometry 24** (1986) 97-132.

[40] Y.S. Poon,"On the Algebraic Dimension of Twistor Spaces," **Math. Ann. 282** (88) 621-627.

[41] Y.S. Poon, "Algebraic Structure of Twistor Spaces," *preprint*, 1990.

[42] W. Thurston, "Three-Dimensional Manifolds, Kleinian Groups, and Hyperbolic Geometry," **Bull. AMS 6** (1982) 357-381.

[43] K.P. Tod,"Compact 3-Dimensional Einstein-Weyl Structures," *preprint*, 1990.

[44] R. S. Ward, "Einstein-Weyl Spaces and **SU**(∞) Toda Fields," **Class. Quantum Grav. 7** (1990) L95-L98.

[45] S. T. Yau, "On the Curvature of Compact Hermitian Manifolds," **Inv. Math. 25** (1974) 213-239.

[46] S. T. Yau, "On Calabi's Conjecture and Some New Results in Algebraic Geometry," **Proc. Nat. Acad. Sci. USA 74** (1977) 1798-1799.

11

Stability and Einstein-Kähler Metric of a Quartic del Pezzo Surface

Dedicated to Professor Tadashi Nagano on his sixtieth birthday

TOSHIKI MABUCHI* AND SHIGERU MUKAI**

* College of General Education, Osaka University, Toyonaka, Osaka, 560 Japan

**Department of Mathcmatics, Nagoya University, Chikusa, Nagoya, 464-01 Japan

Introduction

A *del Pezzo orbifold* S is a compact irreducible complex surface, with only isolated quotient singularities, such that $c_1(S) > 0$. The self-intersection number $c_1(S)^2$ is called its degree, and a smooth S is called a *del Pezzo surface*. Let Σ be the set of all isomorphism classes of quartic del Pezzo surfaces. In this paper, we study two geometric compactifications $\bar{\Sigma}_{EK}$ and $\bar{\Sigma}_{\text{alg}}$ of Σ and show that they coincide.

For a fixed total volume, any quartic del Pezzo surface has a unique Einstein-Kähler metric. Hence Σ carries the Hausdorff distance. By Gromov's precompactness theorem (cf. [**6**],[**8**]), any sequence $\{S_i\}$ in Σ always contains a Cauchy subsequence converging to a metric space S_∞ with a complex structure. Moreover, S_∞ has a natural structure of an Einstein-Kähler orbifold (cf. [**1**],[**2**],[**4**]). The first compactification $\bar{\Sigma}_{EK}$ above is the set of all such limits S_∞ modulo holomorphic isometries.

Let S be a quartic del Pezzo orbifold with only rational double points. Then S has five linearly independent anticanonical forms which allow us to embed S into $\mathbb{P}^4$. The anticanonical image is an intersection of two hyperquadrics $q = 0$ and

*Supported partially by the Max-Planck-Institut für Mathematik, Bonn.

$r = 0$. The point of the Grassmannian $G(2, \mathrm{S}^2(\mathbb{C}^5))$ corresponding to the complex 2-plane spanned by q and r in $\mathrm{S}^2(\mathbb{C}^5)$ is called the (quadratic) Hilbert point of S. If S is nonsingular, then $S \subset \mathbb{P}^4$ is *diagonalizable*, i.e., two quadratic forms q and r defining S are simultaneously diagonalized for a suitable choice of a basis of $\mathbb{C}^5$. Moreover, the Hilbert point is stable with respect to the action of SL(5) if and only if S is nonsingular. Our second compactification $\bar{\Sigma}_{\mathrm{alg}}$ is the geometric quotient of $G(2, \mathrm{S}^2(\mathbb{C}^5))$ by the action of PGL(5). Its boundary consists of PGL(5)-orbits of weakly stable points. Our main results are the following (see §5 and §6):

Theorem A. *The Einstein-Kähler del Pezzo orbifold S_∞ above has only nodes as its singularity. Furthermore, $\mathcal{O}(K_{S_\infty}^{-1})$ is a very ample invertible sheaf and the anticanonical image of S_∞ in $\mathbb{P}^4$ is diagonalizable.*

Theorem B. *Let S be an intersection of two distinct hyperquadrics $q = r = 0$ in $\mathbb{P}^4$. Then the Hilbert point is weakly stable, i.e.,* SL(5)*-orbit of $q \wedge r$ in $\wedge^2 \mathrm{S}^2(\mathbb{C}^5)$ is closed, if and only if $S \subset \mathbb{P}^4$ is diagonalizable and has only nodes as its singularity.*

By combining these theorems, we have a homeomorphism $\bar{\Sigma}_{EK} \cong \bar{\Sigma}_{\mathrm{alg}}$ between the compactifications which is the identity map when restricted to Σ. In particular, we have (see Theorem 1.2):

Corollary C. *A diagonalizable quartic del Pezzo orbifold with only nodes admits an Einstein-Kähler metric.*

Let V_5 be a quintic del Pezzo surface. Then V_5 is the blowing-up of $\mathbb{P}^2$ at four points in a general position, and the group $\mathrm{Aut}(V_5)$ of holomorphic automorphisms of V_5 is isomorphic to the symmetric group $\mathfrak{S}_5$. There are ten exceptional curves of the first kind ℓ_{ij} $(1 \leq i < j \leq 5)$ on V_5. Let Q_p be the blowing-up of V_5 at a point $p \in V$. If p does not lie on $E = \cup_{i,j} \ell_{ij}$, then Q_p is a quartic del Pezzo surface. Since

$\mathrm{Aut}(Q_p)$ acts transitively on the set of exceptional curves of the first kind on Q_p (see Proposition 6.6), the correspondence $V_5 \setminus E \ni p \mapsto Q_p \in \Sigma$ induces an isomorphism $(V_5 \setminus E)/\mathfrak{S}_5 \cong \Sigma$. If p lies on E, then the anticanonical model $\hat{Q}_p \subset \mathbb{P}^4$ of Q_p is a quartic del Pezzo orbifold with only nodes and its Hilbert point is still semistable (see Theorem 6.1). Hence, denoting by $[\pi_p]$ the natural image in $\bar{\Sigma}_{\text{alg}}$ of the semistable Hilbert point, we have

Corollary D. *The correspondence* $V_5 \ni p \mapsto [\pi_p] \in \bar{\Sigma}_{\text{alg}}$ *induces an isomorphism* $V_5/\mathfrak{S}_5 \cong \bar{\Sigma}_{\text{alg}}$.

We here observe that $\hat{Q}_p \subset \mathbb{P}^4$ with $p \in E$ is not diagonalizable. If p is in an intersection of two exceptional curves, $\hat{Q}_p$ is written as the intersection of two hyperquadrics

$$z_0 z_1 + z_2^2 + z_3^2 = z_1^2 + z_2^2 + z_3 z_4 = 0,$$

with two nodes at $(1:0:0:0:0)$ and $(0:0:0:0:1)$. Then the projective transformation $\varphi_{a,b}\colon (z_0 : z_1 : z_2 : z_3 : z_4) \mapsto (az_0 : z_1/a : z_2 : z_3/b : bz_4)$ maps $\hat{Q}_p$ onto the intersection of hyperquadrics

$$z_0 z_1 + z_2^2 + b^2 z_3^2 = a^2 z_1^2 + z_2^2 + z_3 z_4 = 0.$$

Hence, as a and b tend to 0, the image $\varphi_{a,b}(\hat{Q}_p)$ has limit $z_0 z_1 + z_2^2 = z_2^2 + z_3 z_4 = 0$. This limit is diagonalizable, having $\{\varphi_{a,b}\}_{a,b \in \mathbb{C}^*}$ as its automorphisms, and carries an Einstein-Kähler metric. In fact, it is isomorphic to the quotient of $\mathbb{P}^1 \times \mathbb{P}^1$ by an involution (cf. §1). We finally remark that a weakly stable quartic del Pezzo orbifold S has an even number $n = 0, 2, 4$ of nodes and $\dim \mathrm{Aut}(S) = n/2$.

Parts of this note come out of stimulating discussions at Sanda. We wish to thank the Taniguchi Foundation for promoting the symposium at Sanda.

§1. Diagonalizable quartic del Pezzo orbifolds

For $\lambda = (\lambda^{(0)}, \lambda^{(1)}, \dots, \lambda^{(4)}) \in \mathbb{C}^5$, we set $P_\lambda(x) := \Pi_{\alpha=0}^4 (x - \lambda^{(\alpha)})$, and consider the polynomial $P_\lambda(x)$ in x. Let Λ_0, Λ_1, Λ_2 denote the sets of all $\lambda \in \mathbb{C}^5$ such that the equation $P_\lambda(x) = 0$ has only simple roots, just three simple roots and one double root, one simple root and two double roots, respectively. We now put $\Lambda = \Lambda_0 \cup \Lambda_1 \cup \Lambda_2$. Then Λ is nothing but the set of all $\lambda \in \mathbb{C}^5$ such that $P_\lambda(x) = 0$ has only roots of multiplicity ≤ 2. For each $\lambda \in \Lambda$, let S_λ denote the diagonalizable quartic del Pezzo orbifold, with only nodes, defined by

$$S_\lambda := \{ (z_0 : z_1 : \cdots : z_4) \in \mathbb{P}^4 \, ; \, \Sigma_{\alpha=0}^4 z_\alpha^2 = \Sigma_{\alpha=0}^4 \lambda^{(\alpha)} z_\alpha^2 = 0 \}. \tag{1.1}$$

Then, for a suitable choice of a basis of $\mathbb{C}^5$, a diagonalizable quartic del Pezzo orbifold with only nodes is written as S_λ for some $\lambda \in \Lambda$. Note that S_λ is smooth if $\lambda \in \Lambda_0$. Conversely, every quartic del Pezzo surface is isomorphic to S_λ for some $\lambda \in \Lambda_0$. If $\lambda \in \Lambda_1$, then the singularity of S_λ consists of just two A_1-singular points. For instance, if $\lambda \in \Lambda_1$ is such that $\lambda^{(0)} = \lambda^{(1)}$, then the singular points of S_λ are

$$(1 : \sqrt{-1} : 0 : 0 : 0), \qquad (\sqrt{-1} : 1 : 0 : 0 : 0).$$

If $\lambda \in \Lambda_2$, then by setting $W = \{ (z_0 : z_1 : \cdots : z_4) \in \mathbb{P}^4 \, ; \, z_0 z_1 = z_2^2 = z_3 z_4 \}$, we have an isomorphism $S_\lambda \cong W$, where the singularity of W consists of just four A_1-singular points

$$(1 : 0 : 0 : 0 : 0), \quad (0 : 1 : 0 : 0 : 0), \quad (0 : 0 : 0 : 1 : 0), \quad (0 : 0 : 0 : 0 : 1).$$

There exists a double cover of $\mathbb{P}^1 \times \mathbb{P}^1 = \{ (v : w : x : y) \in \mathbb{P}^4 \, ; \, vw = xy \}$ onto W defined by

$$\mathbb{P}^1 \times \mathbb{P}^1 \ni (v : w : x : y) \mapsto (v^2 : w^2 : vw : x^2 : y^2) \in W,$$

which is unramified over the set of regular points of W. For the involution ι sending $(v : w : x : y)$ to $(v : w : -x : -y)$ in $\mathbb{P}^1 \times \mathbb{P}^1$, we have $W = \mathbb{P}^1 \times \mathbb{P}^1/\{1, \iota\}$.

If $\lambda \in \Lambda_0$, then by Tian and Yau [**18**], S_λ admits an Einstein-Kähler metric. For $\lambda \in \Lambda_2$, the surface S_λ admits an Einstein-Kähler metric induced from an Einstein-Kähler metric on $\mathbb{P}^1 \times \mathbb{P}^1$. More generally,

Theorem 1.2. *S_λ admits an Einstein-Kähler metric for all $\lambda \in \Lambda$.*

This is an immediate consequence of Theorem 2.3 (see §2) which will be proved in §5. We here give an outline of another proof for this based on the fact that the methods of Nadel [**13**] are applicable also to the case of orbifolds:

For the proof of Theorem 1.2, it suffices to show the existence of an Einstein-Kähler metric on S_λ for $\lambda \in \Lambda_1$. (The arguments below are valid also for $\lambda \in \Lambda_0$.) Consider the group $G = (\mathbb{Z}_2)^4$ acting on S_λ generated by multiplications of ± 1 on homogeneous coordinates z_i, $1 \leq i \leq 4$, of $\mathbb{P}^4$. Let $\lambda \in \Lambda_1$ and suppose, for contradiction, S_λ admits no Einstein-Kähler metrics. As in nonsingular cases, we have a multiplier subscheme Z of S_λ, preserved by the action of G. Then $\dim_{\mathbb{C}} Z \leq 1$, and moreover, as in [**13**;Corollary 2.5], we obtain

$$H^i(Z, \mathcal{O}_Z) = \mathbb{C} \text{ or } \{0\}, \qquad \text{according as } i = 0 \text{ or } i > 0.$$

Since every G-orbit in S_λ has cardinality ≥ 2, the identity $H^0(Z, \mathcal{O}_Z) = \mathbb{C}$ implies that Z contains a one-dimensional component. Then by $H^1(Z_{\text{red}}, \mathcal{O}_{Z_{\text{red}}}) = \{0\}$, the reduced scheme Z_{red} has a component which is isomorphic to $\mathbb{P}^1$ and preserved by the action of G. Again by the fact that every G-orbit in S_λ has cardinality ≥ 2, we have a nontrivial homomorphic image $\bar{G}$ of G in $\text{Aut}(\mathbb{P}^1)$. Finally, by the classification of finite subgroups of $\text{Aut}(\mathbb{P}^1)$, $\bar{G}$ must be cyclic, hence its action would have a fixed point in contradiction, as required. ∎

§2. Reduction of the problem

Throughout this paper, fix an arbitrary point μ in Λ. Endow Λ with the natural Euclidean distance defined by

$$\mathrm{dist}(\lambda, \nu) := \{ \Sigma_{\alpha=0}^{4} | \lambda^{(\alpha)} - \nu^{(\alpha)} |^2 \}^{1/2},$$

for $\lambda = (\lambda^{(0)}, \ldots, \lambda^{(4)})$ and $\nu = (\nu^{(0)}, \ldots, \nu^{(4)})$ in Λ. Next, define a distance on the space H of all (isometric classes of) compact metric spaces as follows. For compact metric spaces X, Y, with distance functions d_X, d_Y, let $f\colon X \to Y$ be a map which is not necessarily continuous. We call f an *ε-Hausdorff approximation* if (cf. [6],[8])

(a) the ε-neighbourhood of f(X) contains Y, and

(b) $| d_X(p,q) - d_Y(f(p), f(q)) | < \varepsilon$, for $p, q \in X$.

The *Hausdorff distance* $d_H(X, Y)$ between X and Y is the infimum of all positive real numbers ε such that there exist ε-Hausdorff approximations from X to Y and from Y to X. Then d_H defines a distance on the space H.

Let $\{\lambda_j\}_{j=1,2,\ldots}$ be a sequence of points in Λ converging to μ. For each j, the del Pezzo orbifold S_{λ_j} will be denoted simply by S_j, and is assumed to carry an Einstein-Kähler form ω_j in the class $2\pi c_1(S_j)_{\mathbf{R}}$. Let $[S_j, \omega_j] \in H$ denote the metric-space structure associated with the Einstein-Kähler orbifold (S_j, ω_j). Put $X_j := [S_j, \omega_j]$ for simplicity. Assume that

$$\lambda_j \in \Lambda_0 \ \text{ for all } j. \tag{2.1}$$

Then ω_j above is the unique Einstein-Kähler form on S_j in the class $2\pi c_1(S_j)_{\mathbf{R}}$. Moreover, by Gromov's precompactness theorem (cf. [6]), $\{X_j\}_{j=1,2,\ldots}$ has a convergence subsequence in H. Let Y_∞ be its limit, and assume the following simplified situation:

$$d_H(X_j, Y_\infty) \to 0, \ \text{ as } j \to \infty. \tag{2.2}$$

Then by Anderson [**1**], Bando, Kasue and Nakajima [**4**] (see also [**2**]), there exists a compact connected complex orbifold S_∞ carrying an Einstein-Kähler form ω_∞ in the class $2\pi c_1(S_\infty)_{\mathbf{R}}$ such that we can express $Y_\infty = [S_\infty, \omega_\infty]$. Then a crucial step in the proof of Theorem A is to show the following:

Theorem 2.3. *S_∞ is biholomorphic to S_μ, and in particular S_μ admits an Einstein-Kähler metric.*

This will be proved in §5. Since μ is arbitrary in Λ, it in particular provides a proof for Theorem 1.2. Hence, until the end of this section, assume Theorems 1.2 and 2.3. Taking an Einstein-Kähler form ω_μ on S_μ in the class $2\pi c_1(S_\mu)_{\mathbf{R}}$, we put $X_\infty := [S_\mu, \omega_\mu] \in H$. By the uniqueness of Einstein-Kähler metrics for Fano orbifolds (see Bando and Mabuchi [**3**], Nakagawa [**14**]), Theorem 2.3 above is interpreted as the identity $X_\infty = Y_\infty$ in H. Moreover, it allows us to obtain:

Corollary 2.4. *Even if neither* (2.1) *nor* (2.2) *is assumed, we still have the convergence $d_H(X_j, X_\infty) \to 0$ as $j \to \infty$.*

Proof. Suppose, for contradiction, $d_H(X_j, X_\infty)$ does not converge to 0. Replacing $\{\lambda_j\}$ by its subsequence if necessary, we may assume

$$d_H(X_j, X_\infty) \geq \varepsilon, \qquad j = 1, 2, \ldots, \tag{2.4.1}$$

for some constant $\varepsilon > 0$ independent of j. For each j, take a sequence $\{\lambda_{j,\gamma}\}_{\gamma=1,2,\ldots}$ in Λ_0 such that $\lambda_{j,\gamma} \to \lambda_j$ $(\gamma \to \infty)$. For brevity, $S_{\lambda_{j,\gamma}}$ will be denoted by $S_{j,\gamma}$. Note that $S_{j,\gamma}$ carries a unique Einstein-Kähler form $\omega_{j,\gamma}$ in the class $2\pi c_1(S_{j,\gamma})$. Put $X_{j,\gamma} := [S_{j,\gamma}, \omega_{j,\gamma}] \in H$. By [**2**], replacing $\{\lambda_{j,\gamma}\}_{\gamma=1,2,\ldots}$ by its subsequence if necessary, we have an Einstein-Kähler Fano orbifold $Y_j \in H$ such that $d_H(X_{j,\gamma}, Y_j) \to 0$ $(\gamma \to \infty)$.

Then by Theorem 2.3, $X_j = Y_j$ in H. Hence, for each j, there exists a γ_j such that

$$\text{dist}(\lambda_{j,\gamma_j}, \lambda_j) < 1/j; \tag{2.4.2}$$

$$d_H(X_{j,\gamma_j}, X_j) < 1/j. \tag{2.4.3}$$

Since λ_j converges to μ in Λ, the inequality (2.4.2) yields the convergence $\lambda_{j,\gamma_j} \to \mu$ in Λ as $j \to \infty$. Again by [2], replacing $\{\lambda_j\}$ by its subsequence if necessary, we have an Einstein-Kähler Fano orbifold $Y \in H$ such that

$$d_H(X_{j,\gamma_j}, Y) \to 0, \qquad \text{as } j \to \infty. \tag{2.4.4}$$

Then by Theorem 2.3, $X_\infty = Y$ in H. Hence, by (2.4.3) and (2.4.4), $d_H(X_j, X_\infty)$ converges to 0 in contradiction to (2.4.1), as required. ∎

§3. A meromorphic map of a del Pezzo orbifold to a curve

Let $S = \cup_{i=1}^r U_i$ be a del Pezzo orbifold, where each open set U_i is written as a quotient $\tilde{U}_i/G_i$ of a uniformizing complex open 2-ball $\tilde{U}_i$ by a biholomorphic action of a finite group G_i. The corresponding natural projection will be denoted by $\text{pr}_i \colon \tilde{U}_i \to U_i \subset S$. Since S has only isolated quotient singularities, the singular points of S are $\{u_1, u_2, \ldots, u_n\}$ for some n with $0 \le n \le r$ such that

(a) $\tilde{U}_i = U_i$, and $G_i = \{e\}$, $\quad i > n$;

(b) $G_i (\neq \{e\})$ acts freely on $\tilde{U}_i \setminus \{\tilde{u}_i\}$ $\quad i \le n$,

where each $\tilde{u}_i \in \tilde{U}_i$, $i \le n$, is the only fixed point of the G_i-action on $\tilde{U}_i$ associated to the singular point u_i, and if $n = 0$, then S is understood to be smooth. Put

$S_{\mathrm{reg}} := S \setminus \{u_1, u_2, \ldots, u_n\}$. For a coherent sheaf $\mathcal{F}$ on $\tilde{U}_i$, regard its G_i-invariant subsheaf $(\mathcal{F})^{G_i}$ as a sheaf on U_i. Then by setting locally

$$\Omega^p_{U_i} := (\Omega^p_{\tilde{U}_i})^{G_i} \text{ and } \mathcal{O}(K^m_{U_i}) = (\mathcal{O}(K^m_{\tilde{U}_i}))^{G_i}, \qquad 0 \le p \in \mathbb{Z},\ m \in \mathbb{Z},$$

we have global complex analytic sheaves Ω^p_S, $\mathcal{O}(K^m_S)$ on the orbifold S. In terms of the natural inclusion $\iota\colon S_{\mathrm{reg}} \hookrightarrow S$, we can write

$$\Omega^p_S = \iota_*\Omega^p_{S_{\mathrm{reg}}} \text{ and } \mathcal{O}(K^m_S) = \iota_*\{\,\mathcal{O}(K^m_{S_{\mathrm{reg}}})\,\}.$$

For instance, the sheaf $\mathcal{O}(K_S^{-\ell})$ is an ample invertible sheaf on S for the least common multiple $\ell > 0$ of the orders $|G_i|$, $i = 1,2, \ldots,$ of the groups G_i.

Theorem 3.1. *Let $r \ge 1$ and let L be a linear subspace of $\Gamma(S, \mathcal{O}(K_S^{-1}))$ of dimension $r+1$. Assume that $c_1(S)^2 < 8$. Choose a basis $\{\sigma_0, \sigma_1, \ldots, \sigma_r\}$ for L and suppose that the meromorphic map*

$$\Psi\colon S \to \mathbb{P}^r, \qquad x \mapsto (\sigma_0(x) : \sigma_1(x) : \cdots : \sigma_r(x)),$$

takes S generically surjectively to a curve C in $\mathbb{P}^r$. Then,

(3.1.1) *C is rational;*

(3.1.2) *If S admits an Einstein-Kähler metric, then $r \le 2$.*

Proof. Recall that all singular points of the del Pezzo orbifold S are rational. Hence, S is rational (see for instance Sakai [**15**]) and so is C. This proves (3.1.1). To see (3.1.2), let $r \ge 3$ and assume that S admits an Einstein-Kähler metric. We shall derive a contradiction as follows. For the curve C in $\mathbb{P}^r = \{\,(z_0 : z_1 : \cdots : z_r)\,\}$, we take a general hyperplane section $\{\Sigma^r_{\alpha=0}\,\xi_\alpha z_\alpha = 0\} \cap C$ which is a finite number of distinct nonsingular points $x_1, x_2, \ldots, x_d$ of C. Since C is not contained in any hyperplane of $\mathbb{P}^r$, we have $d = \deg C \geq r \geq 3$. Consider the linear system $\{\operatorname{div}(\sigma)\,;\, 0 \neq \sigma \in L\}$, where $\operatorname{div}(\sigma)$ denotes the Weil divisor on S defined by σ. Then by setting $\tau = \Sigma_{\alpha=0}^{r} \xi_\alpha \sigma_\alpha$, we have

$$\operatorname{div}(\tau) = \Sigma_{k=1}^{d} \Psi^{-1}(x_k) + F,$$

where F is the fixed part and each $\Psi^{-1}(x_k)$ denotes the divisor on S obtained as the meromorphic inverse image of x_k under Ψ. Let $\nu\colon \mathbb{P}^1 \to C$ be the normalization of C. For the points $y_k := \nu^{-1}(x_k)$, $k = 1, 2, \ldots, d$, we choose a meromorphic 1-form η on $\mathbb{P}^1$ having poles, only at y_1 and y_2, of multiplicity one. Regard $\Gamma(\mathbb{P}^1, \mathcal{O}(\Sigma_{k=3}^{d} y_k))$ as the space of all meromorphic functions φ on $\mathbb{P}^1$ such that $\operatorname{div}(\varphi) + \Sigma_{k=3}^{d} y_k$ is an effective divisor. Then

$$\tau \cdot (\nu^{-1} \circ \Psi)^*(\varphi\eta) \in \Gamma(S, \mathcal{O}(K_S^{-1}) \otimes \Omega_S^1) = \Gamma(S, \mathcal{O}(T_S)),$$

for every $\varphi \in \Gamma(\mathbb{P}^1, \mathcal{O}(\Sigma_{k=3}^{d} y_k))$. By our construction, each $\tau \cdot (\nu^{-1} \circ \Psi)^*(\varphi\eta)$ is a vector field on S tangent to the fibres of Ψ. Let $\operatorname{Aut}(S/C)$ denote the group of all biholomorphisms of S preserving the fibres of Ψ. Then

$$\dim \operatorname{Aut}(S/C) \geq \dim \Gamma(\mathbb{P}^1, \mathcal{O}(\Sigma_{k=3}^{d} y_k)) = d - 1 \geq 2.$$

Since Matsushima's theorem [**11**] is valid also for Einstein-Kähler orbifolds, the group $\operatorname{Aut}(S)$ is reductive. Therefore, either the subgroup $\operatorname{Aut}(S/C)$ of $\operatorname{Aut}(S)$ is a reductive algebraic group, or it is nonreductive and $\operatorname{Aut}(S) \geq 3$. Hence, $\operatorname{Aut}(S)$ contains an algebraic subgroup G satisfying one of the following:

(Case 1) $\mathbb{G}_m^2 \cong G \subset \operatorname{Aut}(S/C)$;

(Case 2) $G \cong \mathbb{G}_m^3$;

(Case 3) G is isogenous to $\mathrm{SL}(2)$.

If either Case 1 or Case 2 occurs, then G has only countably many algebraic subgroups. On the other hand, in both cases, we have an isotropy subgroup G_s ($\subset G$) of positive dimension at each point s of S. Hence, G_s is nontrivial and independent of s, in contradiction to the effectiveness of the G-action on S. If Case 3 occurs, then SL(2) acts holomorphically and nontrivially on S. Let $\sigma\colon \tilde{S} \to S$ be the SL(2)-equivariant desingularization of S. Then by **[9]**, $\tilde{S}$ is biholomorphic to one of the following:

(a) Hirzebruch surfaces F_n, $n = 1, 2, \ldots$;

(b) $\mathbb{P}^2$;

(c) $\mathbb{P}^1 \times B$,

where B is a nonsingular irreducible projective curve. We first assume that (b) or (c) is the case. Then every effective divisor on S is numerically semipositive, so that S coincides with $\tilde{S}$. By $c_1(S) > 0$, it now follows that S is isomorphic to $\mathbb{P}^2$ or $\mathbb{P}^1 \times \mathbb{P}^1$, which contradicts $c_1(S)^2 < 8$. Finally, assume that (a) is the case. Then S is obtained from F_n, for some $n \geq 1$, by blowing down a unique rational curve of self-intersection $-n$. Hence, Aut(S) is isomorphic to the non-reductive group Aut(F_n) in contradiction. ∎

Remark 3.2. The above (3.1.1) can be proved by the following general fact: For a compact complex $\mathbb{Q}$-Fano orbifold M, we have $\Gamma(M, \Omega^p_M) = \{0\}$, $p > 0$. To see this fact, take an arbitrary $\theta \in \Gamma(M, \Omega^p_M)$. Since M has a structure of a Kähler orbifold, we have $\partial\theta = \bar{\partial}\theta = 0$, hence $\bar{\theta}$ is regarded as a harmonic $(0,p)$-form. Note that the Kodaira vanishing theorem is valid also for orbifolds. Thus, we obtain $\theta = 0$.

§4. The limit anti-canonical map

Let (S_j, ω_j), $j = 1, 2, \ldots$, be as in §2, assuming (2.1) and (2.2). Retain the same notation as in the preceding sections. For each j, we choose a $\mathbb{C}$-basis

$\{e_{j,0}, e_{j,1}, \ldots, e_{j,4}\}$ for $\Gamma(S_j, \mathcal{O}(K_{S_j}^{-1}))$ satisfying

$$\sum_{\alpha=0}^{4} e_{j,\alpha}^2 = \sum_{\alpha=0}^{4} \lambda_j^{(\alpha)} e_{j,\alpha}^2 = 0 \tag{4.1}$$

in $\Gamma(S_j, \mathcal{O}(K_{S_j}^{-2}))$ in such a way that the corresponding anti-canonical map

$$\Phi_j \colon S_j \hookrightarrow \mathbb{P}^4, \qquad p \mapsto (e_{j,0}(p) : e_{j,1}(p) : \cdots : e_{j,4}(p))$$

naturally embeds S_j in $\mathbb{P}^4 = \{(z_0 : z_1 : \cdots : z_4)\}$ (cf. (1.1)). Let $Z = \{p_1, p_2, \ldots, p_m\}$ denote the set of all singular points in S_∞. In view of [**2**] (see also [**1**],[**4**]), replacing $\{\lambda_j\}$ by its subsequence if necessary, we have an into-diffeomorphism $\varphi_j \colon S_\infty \backslash Z \hookrightarrow S_j$ and the following C^∞ convergences on $S_\infty \setminus Z$:

$$\tilde{\omega}_j := \varphi_j^* \omega_j \to \omega_\infty, \quad \text{as } j \to \infty;$$
$$\tilde{J}_j := \varphi_j^* J_j \to J_\infty, \quad \text{as } j \to \infty,$$

where J_j and J_∞ denote respectively the complex structures of S_j and S_∞. We now set $\Gamma_j := \Gamma(S_j, \mathcal{O}(K_{S_j}^{-1}))$, and consider its pointwise Hermitian inner product $(\ ,\ )_{\omega_j} \colon \Gamma_j \times \Gamma_j \to C^\infty_{\mathbb{C}}(S_j)$ induced by the Kähler metric ω_j. Further by $\tilde{S}_j$, we mean the manifold $S_\infty \setminus Z$ with complex structure $\tilde{J}_j$. We now define

$$N_{j,\alpha} := \left\{ \int_{S_j} (e_{j,\alpha}, e_{j,\alpha})_{\omega_j} \omega_j^2/2! \right\}^{1/2};$$
$$\tilde{e}_{j,\alpha} := \varphi_j^*(e_{j,\alpha}/N_{j,\alpha}) \in \Gamma(\tilde{S}_j, \mathcal{O}(K_{\tilde{S}_j}^{-1})).$$

Let $\wedge^2 \mathcal{T}$ denote the sheaf of germs of complex-valued C^∞ skew-symmetric contravariant tensor fields on $S_\infty \setminus Z$. Then by the natural inclusion

$$\Gamma(\tilde{S}_j, \mathcal{O}(K_{\tilde{S}_j}^{-1})) \hookrightarrow \Gamma(S_\infty \setminus Z, \wedge^2 \mathcal{T}),$$

we regard each $\tilde{e}_{j,\alpha}$ as an element of $\Gamma(S_\infty \setminus Z, \wedge^2 \mathcal{T})$. Replacing $\{X_j\}$ by its subsequence if necessary, we have $e_\alpha \in \Gamma(S_\infty, \mathcal{O}(K_{S_\infty}^{-1}))$ such that (see Tian [**17**])

$$\tilde{e}_{j,\alpha} = \varphi_j^*(e_{j,\alpha}/N_{j,\alpha}) \to e_\alpha \text{ in } \Gamma(S_\infty \setminus Z, \wedge^2 \mathcal{T}), \quad \text{as } j \to \infty,$$

in C^∞ topology. Consider the group $G = \mathbb{Z}^4$ acting biholomorphically on S_j generated by multiplications of ± 1 on homogeneous coordinates z_j, $1 \le j \le 4$, of $\mathbb{P}^4$. If $\alpha \neq \beta$, there exists an element g of G such that $g_* e_{j,\alpha} = e_{j,\alpha}$ and $g_* e_{j,\beta} = -e_{j,\beta}$. Since G acts isometrically on (S_j, ω_j), it follows that

$$\int_{S_j} (e_{j,\alpha}/N_{j,\alpha}, e_{j,\beta}/N_{j,\beta})_{\omega_j} \omega_j^2/2! = \delta_{\alpha\beta},$$

where $\delta_{\alpha\beta}$ denotes 1 or 0 according as $\alpha = \beta$ or $\alpha \neq \beta$. Then by passing to the limit, we have (cf. [**17**]):

$$\int_{S_\infty} (e_\alpha, e_\beta)_{\omega_\infty} \omega_\infty^2/2! = \delta_{\alpha\beta}. \tag{4.2}$$

Interchanging the coordinates of $\mathbb{C}^5 = \{(\lambda^{(0)}, \lambda^{(1)}, \dots, \lambda^{(4)})\}$ (hence we renumber $\{e_{j,0}, e_{j,1}, \dots, e_{j,4}\}$ correspondingly) and replacing $\{X_j\}$ by its subsequence if necessary, we may assume from the beginning that

$$N_{j,0} \ge N_{j,1} \ge \cdots \ge N_{j,4}, \qquad j = 1, 2, \dots, \tag{4.3}$$

$$N_{j,\alpha}/N_{j,0} \to N_\alpha \ (\text{as } j \to \infty),\ 0 \le \alpha \le 4, \tag{4.4}$$

for some $0 \le N_\alpha \in \mathbb{R}$. In view of (4.3) and (4.4) above, $1 = N_0 \ge N_1 \ge \cdots \ge N_4$. Hence, for some $k \in \mathbb{Z}$ with $0 \le k \le 4$, we have $N_\alpha > 0$ or $N_\alpha = 0$, according as $\alpha \le k$ or $\alpha > k$. Write (4.1) as

$$\sum_{\alpha=0}^{4} (N_{j,\alpha}/N_{j,0})^2 (e_{j,\alpha}/N_{j,\alpha})^2 = \sum_{\alpha=0}^{4} \lambda_j^{(\alpha)} (N_{j,\alpha}/N_{j,0})^2 (e_{j,\alpha}/N_{j,\alpha})^2 = 0.$$

Set $f_\alpha := N_\alpha e_\alpha$ for $0 \le \alpha \le k$, and also set $f_\alpha := e_\alpha$ for $\alpha > k$. Then by passing to the limit, we have

$$\sum_{\alpha=0}^{k} f_\alpha^2 = \sum_{\alpha=0}^{k} \mu^{(\alpha)} f_\alpha^2 = 0. \tag{4.5}$$

Note that, by (4.2), $f_0, f_1, \dots, f_4$ are $\mathbb{C}$-linearly independent. Clearly, we have $k \ge 1$, since otherwise, (4.5) would imply that $f_0 = 0$ on S_∞ in contradiction. We now have a meromorphic map

$$\Phi: S_\infty \to \mathbb{P}^k, \qquad x \mapsto \Phi(x) := (f_0(x) : f_1(x) : \cdots : f_k(x)),$$

whose meromorphic image $\mathcal{I}m\Phi$ is an irreducible variety of positive dimension and is not contained in any hyperplane of $\mathbb{P}^k$. The following cases are possible:

(Case 1) $k = 4$;

(Case 2) $k = 3$, and $\mu^{(0)}, \mu^{(1)}, \dots, \mu^{(3)}$ are all distinct;

(Case 3) $k = 3$, and $\mu^{(\alpha)} = \mu^{(\beta)}$ for some distinct $\alpha, \beta \in \{0, 1, 2, 3\}$;

(Case 4) $k = 2$;

(Case 5) $k = 1$.

We claim that only Case 1 is possible. If Case 2 occurs, then $\mathcal{I}m\Phi$ is just the nonsingular elliptic curve $\{ (z_0 : z_1 : z_2 : z_3) ; \Sigma_{\alpha=0}^3 z_\alpha^2 = \Sigma_{\alpha=0}^3 \mu^{(\alpha)} z_\alpha^2 = 0 \}$ in $\mathbb{P}^3$, in contradiction to Theorem 3.1. If Case 5 occurs, then by (4.5), $\mathcal{I}m\Phi$ is either $(1 : \sqrt{-1})$ or $(1 : -\sqrt{-1})$ in $\mathbb{P}^1$, which is a contradiction. If Case 4 occurs, then by (4.5), $\sum_{\gamma=1}^2 (\mu^{(\gamma)} - \mu^{(0)}) f_\gamma^2 = 0$, hence $\mathcal{I}m\Phi$ sits in a union of two hyperplanes in $\mathbb{P}^2$ in contradiction. Finally, suppose that Case 3 occurs. Put $\xi := \mu^{(\alpha)} = \mu^{(\beta)}$ with $\alpha \ne \beta$ in $\{0, 1, 2, 3\}$. Then by (4.5), $\sum_{\gamma=0}^3 (\mu^{(\gamma)} - \xi) f_\gamma^2 = 0$. Therefore, $\mathcal{I}m\Phi$ again sits in a union of two hyperplanes in contradiction. Thus, only Case 1 can occur and moreover, by Theorem 3.1, we obtain

Theorem 4.6. *The above Φ defines a generically surjective meromorphic map of S_∞ to S_μ.*

§5. Proof of Theorem A

In this section, we shall prove Theorem A. Using the same notation as in §4, we shall first complete the proof of Theorem 2.3 by showing that the generically surjective meromorphic map $\Phi\colon S_\infty \ni x \mapsto \Phi(x) = (f_0(x) : f_1(x) : \ldots : f_4(x)) \in S_\mu$ is actually biholomorphic.

Theorem 5.1. *$\Phi\colon S_\infty \to S_\mu$ is biholomorphic.*

Proof. Let $\Delta\,(\subset S_\infty)$ be the set of points of indeterminacy of Φ. Choose a proper modification $\theta\colon S'_\infty \to S_\infty$ of a nonsingular projective algebraic surface S'_∞ onto S_∞ such that (i) θ maps $S'_\infty \setminus \theta^{-1}(\Delta)$ biholomorphically onto $S_\infty \setminus \Delta$, and that (ii) $\Phi \circ \theta\colon S'_\infty \to S_\mu$ is a holomorphic map (denoted by Φ' for brevity). Let Δ^μ be the subset $\Phi'(\theta^{-1}(\Delta))$ of S_μ. In view of $\Delta^\mu \neq S_\mu$, we choose two general hyperplane sections $H_\beta \cap S_\mu = \operatorname{div}(h_\beta)$, $\beta = 1,2$, on $S_\mu\,(\subset \mathbb{P}^4(\mathbb{C}))$ such that

$$(H_1 \cap H_2 \cap S_\mu) \cap \Delta^\mu = \phi,$$

where each h_β is in $\Sigma_{\alpha=0}^4 \mathbb{C} f_\alpha$. Since the degree of S_μ in $\mathbb{P}^4$ is four, the generality of the above hyperplane sections implies that $H_1 \cap H_2 \cap S_\mu$ consists of just four distinct points. Suppose, for contradiction, that the linear system $\{\operatorname{div}(f)\,;\, 0 \neq f \in \Sigma_{\alpha=0}^4 \mathbb{C} f_\alpha\}$ has a base point. We set

$$\operatorname{div}(h_\beta) = D_\beta + F, \qquad \beta = 1, 2,$$

where $D_\beta := \theta({\Phi'}^{-1}(H_\beta \cap S_\mu))$ is the moving part and F is the fixed part. Note that D_β and F are effective divisors. Then we have one of the following:

(Case 1) $\quad F \neq 0;$

$$\text{(Case 2)} \qquad F = 0 \text{ and } (D_1 \cap D_2) \setminus \theta(\Phi'^{-1}(H_1 \cap H_2 \cap S_\mu)) \neq \phi.$$

Let (,) denote the intersection pairing on S_∞ in terms of Mumford's intersection theory for normal surfaces (see, for instance, [**16**; p.452]). If Case 1 occurs, then

$$4 = c_1(S_\infty)^2 = c_1(S_\infty)[D_2 + F] > c_1(S_\infty)[D_2] = (D_1 + F, D_2) \geq (D_1, D_2) \geq 4,$$

in contradiction. On the other hand, if Case 2 occurs, we again obtain a contradiction from the following computation:

$$4 = c_1(S_\infty)^2 = (D_1, D_2) > \text{Cardinality}\{\, \Phi'^{-1}(H_1 \cap H_2)\,\} \geq 4.$$

Thus, the above linear system has no base points. Hence, $\Phi\colon S_\infty \to S_\mu$ is a surjective holomorphic map. Then Φ must be biholomorphic. Because otherwise, Zariski's Main Theorem would allow us to obtain a fibre $\Phi^{-1}(x)$ of positive dimension and

$$c_1(S_\infty)[\Phi^{-1}(x)] = (\Phi^{-1}(H_1), \Phi^{-1}(x)) = 0$$

in contradiction to the positivity of $c_1(S_\infty)$, as required. ∎

Define a compactification $\hat{\Sigma}$ of Σ (see the introduction) as follows. For complex orbifolds S and S', we write $S \cong S'$ if S and S' are biholomorphic. Moreover, for Kähler orbifolds (S, ω) and (S', ω'), we write $(S, \omega) \cong (S', \omega')$ if there exists a holomorphic isometry between them. In view of §1, we now set

$$\hat{\Sigma}' := \{\, S_\lambda\, ;\, \lambda \in \Lambda \,\}/ \cong,$$
$$\hat{\Sigma}'' := \{\, (S_\lambda, \omega)\, ;\, \lambda \in \Lambda,\, \omega \in \mathcal{E}_\lambda \,\}/ \cong,$$

where $\mathcal{E}_\lambda (\neq \phi)$ is the set of all Einstein-Kähler forms on S_λ representing $2\pi c_1(S_\lambda)_{\mathbf{R}}$. Then by the uniqueness of Einstein-Kähler metrics (see [**3**],[**14**]), the map sending each (S_λ, ω) to S_λ naturally induces a bijection of $\hat{\Sigma}''$ onto $\hat{\Sigma}'$. We can therefore write both $\hat{\Sigma}'$ and $\hat{\Sigma}''$ simply as $\hat{\Sigma}$. Later, $\hat{\Sigma}$ will be shown to have a structure of a compact Hausdorff space.

Let $g = \Sigma_{p=0}^5 c_p \xi^p \eta^{5-p}$ be a binary quintic. Then, associating to each such g the vector $(c_0, c_1, \ldots, c_5) \in \mathbf{C}^6$, we can identify $\mathbf{C}^6$ with the space of binary quintics. Let us now consider the action of SL(2) on $\mathbf{C}^6$ induced naturally by the SL(2)-action on $\mathbf{C}^2 = \{(\xi, \eta)\}$. Let $g \neq 0$. In terms of the SL(2)-action on $\mathbf{C}^6$, we call g *stable* or *semistable*, according as the corresponding point $[g] := (c_0 : c_1 : \cdots : c_5)$ in $\mathbf{P}^5$ is stable or semistable. Recall the following criterion for semistability (see [**12**; p.78]):

Fact 5.2. *For a binary quintic $g = g(\xi, \eta) \neq 0$, the following two conditions are equivalent:*

(1) *The quintic form $g(\xi, \eta)$ has no zeros of multiplicity ≥ 3 on $\mathbf{P}^1 = \{(\xi : \eta)\}$;*

(2) *g is semistable (or equivalently, stable in this case).*

For each $\lambda = (\lambda^{(0)}, \lambda^{(1)}, \ldots, \lambda^{(4)}) \in \mathbf{C}^5$, let A_λ denote the diagonal matrix of order five whose $(i+1)$-th diagonal element, $0 \leq i \leq 4$, is $\lambda^{(i)}$. Moreover, let I_5 be the identity matrix of order five. We then define a binary quintic $g_\lambda = g_\lambda(\xi, \eta)$ by

$$g_\lambda(\xi, \eta) := \det(\xi A_\lambda - \eta I_5).$$

Then Fact 5.2 shows that g_λ is semistable (or equivalently stable in this case) if and only if $\lambda \in \Lambda$ (see §1). Let $\mathcal{Z}_\lambda$ be the zero-cycle on the complex projective space $\mathbf{P}^1 = \{(\xi : \eta)\}$ defined by $g_\lambda(\xi, \eta) = 0$. Then $\mathcal{Z}_\lambda$ is naturally regarded as an element of the symmetric product $(\mathbf{P}^1)^5/\mathfrak{S}_5 \,(= \mathbf{P}^5)$ of $\mathbf{P}^1$. Consider the geometric quotient

$$\mathrm{PGL}(2)\backslash\backslash(\mathbf{P}^1)^5/\mathfrak{S}_5 \,(= \mathrm{PGL}(2)\backslash\backslash\mathbf{P}^5),$$

which is just the compact Hausdorff quotient, by the action of PGL(2) (see [**12**; p.76]), of the set of all semistable (or equivalently stable) points in $(\mathbb{P}^1)^5/\mathfrak{S}_5$. Since every point in $(\mathbb{P}^1)^5 = (\mathbb{C} \cup \{\infty\})^5$ can be transferred to a point in $\mathbb{C}^5$ by the action of PGL(2), the correspondence of $\mathcal{Z}_\lambda$ to S_λ with $\lambda \in \Lambda$ induces a surjection

$$\mathrm{PGL}(2)\backslash\backslash(\mathbb{P}^1)^5/\mathfrak{S}_5 \ni [\mathcal{Z}_\lambda] \mapsto [S_\lambda] \in \hat{\Sigma},$$

which is easily seen to be bijective. Hence, $\hat{\Sigma} = \mathrm{PGL}(2)\backslash\backslash(\mathbb{P}^1)^5/\mathfrak{S}_5$. By this identification, we give a topology on $\hat{\Sigma}$. Then, $\hat{\Sigma}$ is a compact Hausdorff space.

Proof of Theorem A. Let S_i, $i = 1,2,\dots$, be an arbitrary sequence of quartic del Pezzo surfaces. Then by the notation in §1, we can find $\lambda_i \in \Lambda_0$ such that $S_i = S_{\lambda_i}$ for all i. Since $\hat{\Sigma}$ is compact, each λ_i can be chosen in such a way that, replacing $\{\lambda_i\}$ by its subsequence if necessary, we may assume that the sequence $\{\lambda_i\}$ is convergent in Λ. Then Theorem A is straightforward from Theorem 2.3. ∎

Remark 5.3. It is well-known (see for instance [**7**]) that the geometric quotient $\mathrm{PGL}(2)\backslash\backslash(\mathbb{P}^1)^5$ is the quintic del Pezzo surface V_5. Moreover, $\mathfrak{S}_5$ acts on V_5 as the group $\mathrm{Aut}(V_5)$. Therefore, $\hat{\Sigma} = \mathrm{PGL}(2)\backslash\backslash(\mathbb{P}^1)^5/\mathfrak{S}_5 \cong V_5/\mathfrak{S}_5$. In the next section (see Remark 6.9), we shall show that $\hat{\Sigma}$ coincides with the compactification Σ_{alg} in the introduction.

§6. Stability of a pencil of quadrics

Let $q(z) := \Sigma_{0\le\alpha\le\gamma\le 4}\, a_{\alpha\gamma} z_\alpha z_\gamma$ and $r(z) := \Sigma_{0\le\alpha\le\gamma\le 4}\, b_{\alpha\gamma} z_\alpha z_\gamma$ be quadratic forms in five variables and π their wedge product $q \wedge r \in \wedge^2 \mathrm{S}^2(\mathbb{C}^5)$. Then the point $[\pi] \in \mathbb{P}_*(\wedge^2 \mathrm{S}^2(\mathbb{C}^5))$ is called the (quadratic) Hilbert point of the intersection $S : q(z) = r(z) = 0$ in $\mathbb{P}^4$. Recall that $[\pi]$ is *stable* if the SL(5)-orbit through π is closed and the stabilizer group is finite. Moreover, $[\pi]$ is called *semistable* if the closure of $\mathrm{SL}(5)\cdot\pi$ does not contain the origin. In this section, applying the geometric invariant theory to the action of SL(5) on $\wedge^2 \mathrm{S}^2(\mathbb{C}^5)$, we shall show Theorem B and

Theorem 6.1. (1) *The Hilbert point* $[\pi]$ *is stable if and only if* S *is smooth.* (2) *The point* $[\pi]$ *is semistable if and only if* S *has only nodes as its singularity.*

For $(\xi, \eta) \in \mathbb{C}^2$, let $f_\pi(\xi, \eta)$ denote the discriminant $\operatorname{discr}(\xi q(z) + \eta r(z))$ of the quadratic form $\xi q(z) + \eta r(z)$. Then f_π is a binary quintic and the associated orbit $\mathrm{SL}(2) \cdot f_\pi \in \mathrm{S}^5(\mathbb{C}^2)$ depends only on π. We first observe the following:

Proposition 6.2. *For any* $\mathrm{SL}(2)$*-invariant* G *of binary quintics, there exists an* $\mathrm{SL}(5)$*-invariant polynomial* $\bar{G}$ *on* $\wedge^2 \mathrm{S}^2(\mathbb{C}^5)$ *such that the identity* $G(f_\pi) = \bar{G}(\pi)$ *holds for all* $\pi = q \wedge r$.

Proof. We write $f_\pi(\xi, \eta) = \Sigma_{p=0}^5 c_p(a,b) \xi^p \eta^{5-p}$, where coefficients $c_p(a,b)$ are polynomials in thirty variables $a_{\alpha\gamma}$'s and $b_{\alpha\gamma}$'s. By identifying f_π and the associated $(c_0(a,b), \ldots, c_5(a,b))$, we consider the polynomial $\tilde{G}(a,b) := G(c_0(a,b), \ldots, c_5(a,b))$ on $\mathbb{C}^{30}$. Note that G is an SL(2)-invariant of binary quintics, i.e.,

$$\tilde{G}(ua + vb, wa + xb) = \tilde{G}(a,b) \qquad \text{for all} \begin{pmatrix} u & v \\ w & x \end{pmatrix} \in \mathrm{SL}(2).$$

Hence, the first fundamental theorem on vector invariants (see [**19**; p.45]) shows that $\tilde{G}$ is a polynomial in the 2×2 minors of the 2×15 matrix

$$\begin{pmatrix} a \\ b \end{pmatrix} = \begin{pmatrix} \ldots, a_{\alpha\gamma}, \ldots \\ \ldots, b_{\alpha\gamma}, \ldots \end{pmatrix},$$

that is, $\tilde{G}(a,b)$ is written as a polynomial $\bar{G}$ in $a_{\alpha\gamma} b_{\beta\delta} - b_{\alpha\gamma} a_{\beta\delta}$'s. Since the discriminant is SL(5)-invariant, so is the polynomial $\bar{G}$ on $\wedge^2 \mathrm{S}^2(\mathbb{C}^5)$. ∎

Note that f_π is semistable if and only if there exists an SL(2)-invariant G of binary quintics such that $G(f_\pi) \neq 0$. By Fact 5.2 and Proposition 6.2, we obtain

Corollary 6.3. *If $f_\pi = f_\pi(\xi, \eta)$ has no zeros of multiplicity ≥ 3 on $\mathbb{P}^1 = \{(\xi : \eta)\}$, then the Hilbert point $[\pi] \in \mathbb{P}_*(\wedge^2 \mathrm{S}^2(\mathbb{C}^5))$ is semistable.*

Let $\mathcal{P}_\pi$ be the set of all triples (q', r', π') of quadratic forms $q'(z)$, $r'(z)$ and their wedge product $\pi' = q' \wedge r'$ such that the Hilbert point $[\pi']$ is weakly stable and that the binary quintic $f_{\pi'}$ sits in the orbit $\mathrm{SL}(2) \cdot f_\pi$. If f_π is stable (so that its SL(2)-orbit is separated from the others by SL(2)-invariants) and moreover if $[\pi]$ is not weakly stable, then in view of Proposition 6.2, the closure of the orbit $\mathrm{SL}(5) \cdot \pi$ in $\wedge^2 \mathrm{S}^2(\mathbb{C}^5)$ contains a closed orbit $\mathrm{SL}(5) \cdot \pi'$ of smaller dimension for some triple $(q', r', \pi') \in \mathcal{P}_\pi$. We here observe that the orbit $\mathrm{GL}(5) \cdot \pi$ is uniquely determined by $\mathrm{GL}(2) \cdot f_\pi$ whenever two quadratic forms $q(z)$ and $r(z)$ are simultaneously diagonalized for a suitable choice of a basis of $\mathbb{C}^5$. Hence, we have

Corollary 6.4. *Suppose that $S \subset \mathbb{P}^4$ is diagonalizable and that the binary quintic f_π is semistable (or equivalently stable). Then the Hilbert point $[\pi]$ is weakly stable, provided that the intersection $S' : q'(z) = r'(z) = 0$ is diagonalizable in $\mathbb{P}^4$ for all $(q', r', \pi') \in \mathcal{P}_\pi$.*

Now, without any assumption on diagonalizability, we analyze the stability of $[\pi]$ and the singularity of S in terms of the multiplicity of the zeros of f_π on $\mathbb{P}^1$. Then the following two cases are possible:

(Case 1) f_π has no multiple zeros on $\mathbb{P}^1$;

(Case 2) f_π has a multiple zero on $\mathbb{P}^1$.

We first consider Case 1. Let $(\xi_0 : \eta_0)$, $(\xi_1 : \eta_1)$, $\ldots$, $(\xi_4 : \eta_4)$ be the five distinct zeros of f_π and take, for each $0 \leq i \leq 4$, a nonzero vector $v_i \in \mathbb{C}^5$ representing a singular point of the quadric $\xi_i q(z) + \eta_i r(z) = 0$ in $\mathbb{P}^4$. Then

$$\xi_i (v_i, v)_q + \eta_i (v_i, v)_r = 0$$

holds for every $v \in \mathbb{C}^5$, where $(\ ,\ )_q$ and $(\ ,\ )_r$ are the bilinear forms associated to $q(z)$ and $r(z)$, respectively. Since the points $(\xi_i : \eta_i)$, $0 \le i \le 4$, are all distinct, we have $(v_i, v_j)_q = (v_i, v_j)_r = 0$ for all i and j with $i \ne j$. Hence, $v_0, v_1, \ldots, v_4$ are linearly independent. Therefore, choosing $\{v_0, v_1, \ldots, v_4\}$ as a new basis for $\mathbb{C}^5$, we may rewrite the quadratic forms $q(z)$ and $r(z)$ in a normalized form

$$q(z) = -\eta_0 z_0^2 - \eta_1 z_1^2 - \cdots - \eta_4 z_4^2,$$
$$r(z) = \xi_0 z_0^2 + \xi_1 z_1^2 + \cdots + \xi_4 z_4^2.$$

Hence, S is smooth and is a diagonalizable quartic del Pezzo surface in $\mathbb{P}^4$. Now by Corollary 6.4, the Hilbert point $[\pi]$ is weakly stable. Recall that the elementary abelian group E_{16} of order sixteen acts effectively on S by changing the signs of coordinates. The quotient surface S/E_{16} is isomorphic to $\mathbb{P}^2$ and the morphism $\kappa\colon S \to \mathbb{P}^2$ is a Kummer covering whose branch is a union of five lines $\ell_1, \ell_2, \ldots, \ell_5$. Hence $\mathrm{Aut}(S)$ is finite, since $\mathrm{Aut}(S)/E_{16}$ is isomorphic to a subgroup of the symmetric group $\mathfrak{S}_5$ of degree five. In particular, $[\pi]$ is stable. Thus we obtain

Proposition 6.5. *If the binary quintic f_π has no multiple zeros on $\mathbb{P}^1$, then S is diagonalizable and smooth, and moreover, the Hilbert point $[\pi]$ is stable.*

Let $C \subset \mathbb{P}^2$ be an irreducible conic which is tangent to all lines $\ell_1, \ldots, \ell_5$. Then $\kappa^{-1}(C)$ is a union of sixteen lines on $S \subset \mathbb{P}^4$. Since S is nothing but the blowing-up of $\mathbb{P}^2$ at five points in a general position, these sixteen lines are all lines on $S \subset \mathbb{P}^4$. Hence, we have

Proposition 6.6. *The elementary abelian group $E_{16} \subset \mathrm{Aut}(S)$ acts freely on the set of all lines on S in $\mathbb{P}^4$.*

We next assume that Case 2 occurs. In this case, we may assume that $(0:1)$ is a multiple zero. We take a system of homogeneous coordinates in such a way that the quadratic form $r(z)$ is diagonalized. Since the hypersurface $r(z) = 0$ in $\mathbb{P}^4$ is singular, we may write $r(z)$ as $z_0^2 + \cdots + z_k^2$ for some $k \leq 3$. Since $f_\pi(\xi, 1) = \mathrm{discr}(\xi q(z) + r(z))$ is divisible by ξ^2, we have either the inequality $k < 3$ or the equalities $k = 3$ and $a_{44} = 0$. Note that Case 2 is divided into the following two subcases:

(2-a) The point $(0 : 1)$ is a double zero for f_π;

(2-b) The point $(0 : 1)$ is a zero for f_π of multiplicity ≥ 3.

First, assume that (2-a) is the case. Then the pair of quadratic forms $q(z)$ and $r(z)$ satisfies one of the following two conditions:

(i) $k = 2$, and $q(z) = 0$ has two simple zeros on the line $z_0 = z_1 = z_2 = 0$ in $\mathbb{P}^4$;

(ii) $k = 3$, $a_{44} = 0$ and ${a_{04}}^2 + {a_{14}}^2 + {a_{24}}^2 + {a_{34}}^2 \neq 0$.

In the case (i), the homogeneous coordinate system $(z_0, z_1, \ldots, z_4)$ on $\mathbb{P}^4$ can be chosen in such a way that $q(0,0,0,1,0) = q(0,0,0,0,1) = 0$. Therefore, we may write

$$q(z) = q_1(z_0, z_1, z_2) + m_1(z_0, z_1, z_2)z_3 + m_2(z_0, z_1, z_2)z_4 + z_3 z_4,$$

for some quadratic form q_1 and linear forms m_1, m_2 in three variables. Taking $z_3 + m_2(z_0, z_1, z_2)$ and $z_4 + m_1(z_0, z_1, z_2)$ as new coordinates z_3 and z_4, respectively, we may further write the quadratic form $q(z)$ as $q_2(z_0, z_1, z_2) + z_3 z_4$ for some quadratic form q_2 in three variables. Hence, the surface $S : q(z) = r(z) = 0$ carries a $\mathbb{G}_m$-action which fixes two nodes $(0 : 0 : 0 : 0 : 1)$ and $(0 : 0 : 0 : 1 : 0)$ of S.

In the case (ii), take a complex orthogonal transformation for the space $\mathbb{C}z_0 + \cdots + \mathbb{C}z_3$ such that the associated coordinate change allows us to write

$$({a_{04}}^2 + {a_{14}}^2 + {a_{24}}^2 + {a_{34}}^2)^{-1/2} \cdot (a_{04}z_0 + a_{14}z_1 + a_{24}z_2 + a_{34}z_3)$$

as a new coordinate z_3. Moreover, replacing z_4 by its suitable constant multiple, we have $q(z) = q_3(z_0, z_1, z_2, z_3) + z_3 z_4$ for some quadratic form q_3 in four variables. Then the action $(z_3, z_4) \mapsto (tz_3, t^{-1}z_4)$, $t \in \mathbb{C}^*$, of $\mathbb{G}_m$ transforms $q(z)$ and $r(z)$ into

$$q^{(t)}(z) := q_3(z_0, z_1, z_2, tz_3) + z_3 z_4;$$
$$r^{(t)}(z) := z_0^2 + z_1^2 + z_2^2 + t^2 z_3^2.$$

As t tends to 0, we have limits

$$q^{(0)}(z) = q_3(z_0, z_1, z_2, 0) + z_3 z_4;$$
$$r^{(0)}(z) = z_0^2 + z_1^2 + z_2^2.$$

Hence, the SL(5)-orbit through $q \wedge r$ is not closed and its closure contains $q^{(0)} \wedge r^{(0)}$. Note that the pair of $q^{(0)}$ and $r^{(0)}$ satisfies the condition (i) above. Therefore,

Proposition 6.7. *Assume that $(\xi_0 : \eta_0)$ is a double zero for the quintic form $f_\pi = f_\pi(\xi, \eta)$ on $\mathbb{P}^1$. Let Q_0 be the quadric $\xi_0 q(z) + \eta_0 r(z) = 0$ in $\mathbb{P}^4$. Then $[\pi]$ is not stable and S has a node. Moreover, we have one of the following:*

(i) rk $Q_0 = 3$ *and the singular line of Q_0 meets S at two distinct nodes of S;*

(ii) rk $Q_0 = 4$ *and the unique singular point of Q_0 is a node of S.*

In the latter case, $[\pi]$ is not weakly stable, since $\mathbb{P}^4$ admits a nontrivial action $\{ \phi_t \,;\, t \in \mathbb{C}^* \}$ of $\mathbb{G}_m$ such that $\lim_{t\to 0} \phi_t(q \wedge r)$ is associated with the the former case. In the former case, S admits a nontrivial $\mathbb{G}_m$-action which fixes the above two nodes.

We next assume that (2-b) above occurs. Then the pair of quadratic forms $q(z) = \Sigma_{0 \le \alpha \le \gamma \le 4}\, a_{\alpha\gamma} z_\alpha z_\gamma$ and $r(z) = z_0^2 + \cdots + z_k^2$ satisfies one of the following:

(iii) $k \le 1$;

(iv) $k = 2$ and $q(z)$ has a multiple zero on the line $z_0 = z_1 = z_2 = 0$ in $\mathbb{P}^4$;

(v) $k = 3$ and ${a_{04}}^2 + {a_{14}}^2 + {a_{24}}^2 + {a_{34}}^2 = a_{44} = 0$.

In the case (iii), S is either reducible or non-reduced. The one-parameter subgroup

$$\phi_t(z_0, z_1, z_2, z_3, z_4) := (t^3 z_0, t^3 z_1, t^{-2} z_2, t^{-2} z_3, t^{-2} z_4), \qquad t \in \mathbf{C}^*,$$

of SL(5) maps $q(z) \wedge r(z)$ to $t^2 q(t^5 z_0, t^5 z_1, z_2, z_3, z_4) \wedge r(z)$. Hence, $\lim_{t\to 0} \phi_t(q\wedge r) = 0$ and $[\pi]$ is not semistable.

In the case (iv), via a suitable coordinate change of z_3 and z_4, we may assume that $q(0,0,0,z_3,z_4) = cz_3^2$ with $c = 1$ or 0. Then $(0:0:0:0:1)$ is a singular point of S which is not a node. The one-parameter subgroup

$$\phi_t(z_0, z_1, z_2, z_3, z_4) := (t^2 z_0, t^2 z_1, t^2 z_2, t^{-1} z_3, t^{-5} z_4), \qquad t \in \mathbf{C}^*,$$

maps $q(z) \wedge r(z)$ to $q(t^4 z_0, t^4 z_1, t^4 z_2, t z_3, t^{-3} z_4) \wedge r(z)$. Since $q(z)$ does not contain the term z_4^2 or $z_3 z_4$, we have $\lim_{t\to 0} \phi_t(q \wedge r) = 0$ and $[\pi]$ is not semistable.

In the case (v), the tangent space of the quadric $q(z) = 0$ at the point $(0:0:0:0:1)$ is tangential to the tangent cone of the singular quadric $r(z) = 0$, unless the quadric $q(z) = 0$ is singular at the point. Hence, this point is a singular point of S which is not a node. For a suitable choice of $z_0, \ldots, z_4$, we can write the quadratic forms $q(z)$ and $r(z)$ as

$$q(z) = q_4(z_0, z_1, z_2, z_3) + c z_3 z_4;$$
$$r(z) = z_0^2 + z_1^2 + z_2 z_3$$

for some constant c and a quadratic form q_4 in four variables. Then the one-parameter subgroup

$$\phi_t(z_0, z_1, z_2, z_3, z_4) := (t^2 z_0, t^2 z_1, t^{-1} z_2, t^5 z_3, t^{-8} z_4), \qquad t \in \mathbf{C}^*,$$

of SL(5) maps $q \wedge r$ to $\{ q_4(t^4 z_0, t^4 z_1, t z_2, t^7 z_3) + ct z_3 z_4 \} \wedge (z_0^2 + z_1^2 + z_2 z_3)$. Hence, $\lim_{t\to 0} \phi_t(q\wedge r) = 0$, and $[\pi]$ is not semistable. Thus, in view of Fact 5.2 and Corollary 6.3, we obtain

Proposition 6.8. *If f_π has a zero of multiplicity ≥ 3 on $\mathbb{P}^1$, then $[\pi]$ is not semistable and S has a singular point which is not a node. In particular, f_π is semistable if and only if $[\pi]$ is semistable.*

Proof of Theorem 6.1. By Proposition 6.8 and Fact 5.2, it suffices to consider the case where $[\pi]$ is semistable, i.e., f_π has no zeros of multiplicity ≥ 3 on $\mathbb{P}^1$. To each singular point s_0 of S, we can uniquely associate a zero $(\xi_0 : \eta_0) \in \mathbb{P}^1$ for f_π such that the quadric $\xi_0 q(z) + \eta_0 r(z) = 0$ in $\mathbb{P}^4$ is singular at $z = s_0$. We here claim that $(\xi_0 : \eta_0)$ is a multiple zero for f_π. Note that, by this claim, Propositions 6.5 and 6.7 immediately imply Theorem 6.1. To show this claim, suppose on the contrary that $(\xi_0 : \eta_0)$ is a simple zero for f_π. Now, we may assume $(\xi_0 : \eta_0) = (0 : 1)$ without loss of generality. Then the quadric $r(z) = 0$ is singular, and the coordinates $z_0, \ldots, z_4$ can be chosen in such a way that $r(z)$ is expressible as $z_0^2 + \cdots + z_k^2$ for some $k \leq 3$. Since $f_\pi(\xi, 1) = \operatorname{discr}(\xi q(z) + r(z))$ is not divisible by ξ^2, it follows that $k = 3$ and $a_{44} \neq 0$. Then by $k = 3$, we have $s_0 = (0 : \cdots : 0 : 1)$. Moreover, by $a_{44} \neq 0$, the quadric $q(z) = 0$ does not pass through s_0, which is a contradiction as required. ∎

Proof of Theorem B. By Proposition 6.8 and Fact 5.2, we may assume that both f_π and $[\pi]$ are semistable, i.e., f_π has no zeros of multiplicity ≥ 3 on $\mathbb{P}^1$. Then by (2) of Theorem 6.1, S has only nodes as its singularity. Now by Corollary 6.4, it suffices to show that $S \subset \mathbb{P}^4$ is diagonalizable, with $[\pi]$ assumed to be weakly stable. Let d be the number points on $\mathbb{P}^1$ at which f_π has a double zero. Then we have $0 \leq d \leq 2$. The case $d = 0$ is already proved by Proposition 6.5. We next assume $d = 1$. Since $[\pi]$ is weakly stable, S has just two nodes by Proposition 6.7, and therefore, we can find quadratic forms q' and r' in three variables such that

$$q(z) = q'(z_0, z_1, z_2) + z_3 z_4,$$
$$r(z) = r'(z_0, z_1, z_2),$$

for some suitable choice of $z_0, \ldots, z_4$, where the cubic form $f'(\xi, \eta) := \operatorname{discr}(\xi q' + \eta r')$ has three distinct zeroes in $\mathbb{P}^1$. Then by the same argument as in Case 1 immediately after Corollary 6.4 above, q' and r' are simultaneously diagonalizable and so are q and r. We finally assume $d = 2$. Since $[\pi]$ is weakly stable, S has just four nodes. Then, associated to a double zero for f_π, a suitable choice of system $(z_0, \ldots, z_4)$ allows us to write $q(z)$ and $r(z)$ as

$$q(z) = q'(z_0, z_1, z_2) + z_3 z_4,$$
$$r(z) = r'(z_0, z_1, z_2),$$

for some quadratic forms q' and r' in three variables. Moreover, associated to the other double zero for f_π, we have another basis $\{q'', r''\}$ for $\mathbb{C}q' + \mathbb{C}r'$ such that

$$q''(z_0, z_1, z_2) = z_0 z_1 + c_0 z_2^2,$$
$$r''(z_0, z_1, z_2) = z_2^2,$$

for some constant c_0 with a suitable coordinate change of z_0, z_1 and z_2. Thus, $q(z)$ and $r(z)$ are simultaneously diagonalizable, as required. ∎

Remark 6.9. By §1 and Theorem B, the Hilbert point $[\pi]$ is weakly stable if and only if S is written as S_λ for some $\lambda \in \Lambda$. Therefore, we can identify $\bar{\Sigma}_{\text{alg}}$ and $\hat{\Sigma}$. This gives another proof for Corollary D in the introduction, since we have $\hat{\Sigma} \cong V_5/\mathfrak{S}_5$ by Remark 5.3. On the other hand, §1 and Theorem A show that $\bar{\Sigma}_{EK}$ coincides with $\hat{\Sigma}''(= \hat{\Sigma})$ set-theoretically. For each $\lambda \in \Lambda$, we fix an element ω_λ of $\mathcal{E}_\lambda$, where $\mathcal{E}_\lambda$ is as in §5. Then by Corollary 2.4, as λ tends to μ, the Einstein-Kähler orbifold $(S_\lambda, \omega_\lambda)$ converges to (S_μ, ω_μ) in the Hausdorff distance. In other words, the natural topology of $\hat{\Sigma}(= \bar{\Sigma}_{\text{alg}})$ is compatible with the convergence for $\bar{\Sigma}$ in the Hausdorff distance. Hence, we have a natural homeomorphism between two compact Hausdorff spaces $\bar{\Sigma}_{EK}$ and $\bar{\Sigma}_{\text{alg}}$.

References

[1] M. Anderson, *Ricci curvature bounds and Einstein metrics on compact manifolds*, J. Amer. Math. Soc. **2** (1989), 455–490.

[2] S. Bando, *Bubbling out of Einstein manifolds*, Tôhoku Math. J. **42** (1990), 205–216.

[3] S. Bando and T. Mabuchi, *Uniqueness of Einstein Kähler metrics modulo connected group actions*, in "Algebraic Geometry, Sendai, 1985," Adv. Stud. Pure Math. **10**, Kinokuniya, Tokyo and North-Holland, Amsterdam, 1987, pp. 11–40.

[4] S. Bando, A. Kasue and H. Nakajima, *On a construction of coordinates at infinity on manifolds with fast curvature decay and maximal volume growth*, Invent. Math. **97** (1989), 313–349.

[5] M. Demazure, *Surfaces de Del Pezzo*, in "Séminaire sur les singularités des surfaces," Lecture Notes in Math. **777**, Springer-Verlag, Heidelberg, 1980, pp. 21–69.

[6] K. Fukaya, *Hausdorff convergence of Riemannian manifolds and its applications*, in "Recent Topics in Differential and Analytic Geometry," Adv. Stud. Pure Math. **18-I**, Kinokuniya, Tokyo and Academic Press, Boston, 1990, pp. 143–238.

[7] A. Futaki, *The Ricci curvature of symplectic quotients of Fano manifolds*, Tohoku Math. J. **39** (1987), 329–339.

[8] M. Gromov, "Structures métriques pour les variétés riemanniennes," rédige par J. Lafontaine et P. Pansu, Cedic/Fernand Nathan, Paris, 1981.

[9] T. Mabuchi, *On the classification of essentially effective* $\mathrm{SL}(n;\mathbb{C})$*-actions*, Osaka J. Math. **16** (1979), 745–758.

[10] T. Mabuchi, *Compactification of the moduli space of Einstein-Kähler orbifolds*, in "Recent Topics in Differential and Analytic Geometry," Adv. Stud. Pure Math. **18-I**, Kinokuniya, Tokyo and Academic Press, Boston, 1990, pp. 359–384.

[11] Y. Matsushima, *Sur la structure du groupe d'homéomorphismes analytiques d'une certaine variété kählérienne*, Nagoya Math. J. **11** (1957), 145–150.

[12] D. Mumford and J. Fogarty, "Geometric invariant theory," 2nd edition, Ergebnisse der Math. und ihrer Grenzgebiete **34**, Springer-Verlag, Heidelberg, 1982.

[13] A. M. Nadel, *Multiplier ideal sheaves and Kähler-Einstein metrics of positive scalar curvature*, Annals of Math. **132** (1990), 549–596.

[14] Y. Nakagawa, *An isoperimetric inequality for orbifolds*, in preparation.

[15] F. Sakai, *Anticanonical models of rational surfaces*, Math. Ann. (1984), 389–410.

[16] F. Sakai, *Classification of normal surfaces*, Proc. Symp. Pure Math. **46**, Amer. Math. Soc.,1987, pp.451–465.

[17] G. Tian, *On Calabi's conjecture for complex surfaces with positive first Chern class*, Inventiones Math. **101** (1990), 101–172.

[18] G. Tian and S.-T. Yau, *Kähler-Einstein metrics on complex surfaces with* $C_1 > 0$, Commun. Math. Phys. **112** (1987), 175–203.

[19] H. Weyl, "The classical groups," 2nd edition, Princeton Univ. Press, Princeton, 1946.

12

Geometric Classification of $\mathbb{Z}_2$-Commutative Algebras of Super Differential Operators

MOTOHICO MULASE*

Institute of Theoretical Dynamics
and
Department of Mathematics
University of California
Davis, CA 95616

0. Introduction.

The purpose of this paper is to give a geometric classification of all the supercommutative algebras consisting of super differential operators.

A geometric classification theorem of commutative algebras of ordinary differential operators was established in [**M3**]. The result is, roughly speaking, that there is a bijective correspondence between the set of isomorphism classes of commutative algebras of ordinary differential operators, and the set of isomorphism classes of geometric *quintets* $(C, p, \pi, \mathcal{F}, \phi)$ consisting of an algebraic curve C of an arbitrary genus g, a smooth point $p \in C$, a local covering π of C ramified at p, a semi-stable torsion-free sheaf $\mathcal{F}$ over C of an arbitrary rank r and degree $r(g-1)$, and a local isomorphism ϕ of $\mathcal{F}$ with a free sheaf defined near p. (One can think of ϕ as a local trivialization of $\mathcal{F}$ around p.) Thus a commutative algebra of ordinary differential operators has enormously rich geometric information. All semi-stable torsion-free sheaves satisfying the cohomology vanishing

$$H^0(C, \mathcal{F}) = H^1(C, \mathcal{F}) = 0 \tag{0.1}$$

show up in the above correspondence.

As we have described in the introduction of [**M3**], there is a long history in the study of the problem of classifying commuting ordinary differential operators. The success of solving this historical problem, together with the new trends coming from mathematical physics, promotes generalizations of the theory. There are three different directions one can study:

(1) Classification of *quasicommuting* pairs of ordinary differential operators.
(2) Classification of commutative algebras consisting of partial differential operators.
(3) Classification of supercommutative algebras of super differential operators.

*Research supported in part by NSF Grant DMS 91–03239.

We say two operators P and Q quasicommute if $[P,Q]=1$. The problem (1) of finding all solutions of the equation $[P,Q]=1$ is motivated by the recent research in mathematical physics, in particular, 2-dimensional quantum gravity and matrix models. A geometric classification theorem of the solutions of this equation, sometimes called *string equation*, has been established by Schwarz [**Sc1**] by generalizing the theory of [**M3**]. Surprisingly, the moduli space of the solutions of string equation has a much similar structure of the moduli spaces appearing in the classification of commuting operators.

The second direction, study of commuting partial differential operators, is very difficult. No direct generalization of the theory of ordinary differential operators works. One needs a completely new approach to this problem. Of course there are a lot of examples of commuting partial differential operators, except for the trivial ones like tensor products of ordinary differential operators. These examples come from the theory of *completely integrable systems*. Let

$$L = \frac{d}{dx} + u_2 \left(\frac{d}{dx}\right)^{-1} + u_3 \left(\frac{d}{dx}\right)^{-2} + \cdots$$

be an ordinary pseudodifferential operator with coefficients in $k[[x]]$, where k is an arbitrary field of characteristic zero. We denote by L_+^n the differential operator part of the n-th power of L. The Kadomtsev-Petviashvili system (the KP system) is a system of equations of the form

$$\left[\frac{\partial}{\partial t_m} - L_+^m, \frac{\partial}{\partial t_n} - L_+^n\right] = 0 ,$$

which is nothing but the commutativity condition of partial differential operators. It is shown in [**M1**] that the coefficients u_i of every finite-dimensional solution L are given by the Riemann theta functions associated with a Jacobian variety. Thus the KP system gives a set of commuting partial differential operators defined on a Jacobian variety. Indeed, if the time-evolution parameters are represented on a Jacobian variety by

$$t_j = \sum_i h_{ij} z_i ,$$

where $z_1, z_2, \cdots, z_n$ are coordinates on the Jacobian variety, then

$$\sum_j h_{ij} \left(\frac{\partial}{\partial t_j} - L_+^j\right) , \qquad i = 1, 2, \cdots, n \tag{0.2}$$

gives commuting partial differential operators defined on the Jacobian variety. Therefore, for every n, we can construct a nontrivial example of a set of n commuting partial differential operators. Since these operators are algebraically independent, they form a polynomial ring of n variables, which is nothing but the coordinate ring of the vector space $H^1(C, \mathcal{O}_C)$ of an algebraic curve C. Although the algebra thus obtained from (0.2) is isomorphic to

$$k\left[\frac{\partial}{\partial z_1}, \frac{\partial}{\partial z_2}, \cdots, \frac{\partial}{\partial z_n}\right] ,$$

it is not *equal* to the trivial polynomial ring. From the point of view of the cohomology theory of [**M1**, Section 2], the commuting partial differential operators of (0.2) are in a complementary position to the ring of commuting ordinary differential operators; the latter forms the affine coordinate ring of the algebraic curve C.

Recently, Nakayashiki [**N1**], [**N2**] has started a systematical and very interesting study of commuting partial differential operators which are defined on arbitrary Abelian varieties. At this moment, we still do not know what else will show up on our list of commuting partial differential operators. Further developments are necessary for the final solution to the problem (2).

In this paper, we study supercommutative super differential operators, and solve the problem (3). This problem is also closely related with the study of supersymmetric quantum field theories.

Although supersymmetric generalizations have been established for such theories like the KP theory, the Krichever construction, and the characterization of Jacobian varieties ([**KL**], [**MaR**], [**M2**], [**M4**], [**MR**], [**R**]), nothing has been known until now about supersymmetric counterpart of the theory of geometric classification of commuting ordinary differential operators.

The super Grassmannian of [**M4**] and [**MR**], which was effectively used in the study of Jacobian varieties of supercurves, turns out not to be the right object we need in order to solve the problem (3).

In giving a supersymmetric generalization of the KP theory, Manin and Radul [**MaR**] introduced the notion of super pseudodifferential operators. Let (x, ξ) be a coordinate of the affine superspace $\mathbb{A}_k^{1|1}$ of dimension $1|1$ over k, where x is the usual coordinate of $\mathbb{A}_k^1$ and ξ satisfies $\xi^2 = 0$. A super derivation operator acting on the coordinate ring $k[x, \xi]$ is defined by

$$\delta = \frac{\partial}{\partial \xi} + \xi \frac{\partial}{\partial x} .$$

An expression

$$a_n \delta^n + a_{n-1} \delta^{n-1} + a_{n-2} \delta^{n-2} + \cdots + a_0 + a_{-1} \delta^{-1} + \cdots \tag{0.3}$$

with coefficients in $R = k[[x]] \oplus k[[x]]\xi$ is called a super pseudodifferential operator. If $a_m = 0$ for all negative m, then we call it a super differential operator. Let E denote the set of all super pseudodifferential operators, and D the set of super differential operators. We also use the notation $E^{(-1)}$ for the set of all super pseudodifferential operators (0.3) with $a_m = 0$ for all nonnegative m. Then we have a natural decomposition

$$E = D \oplus E^{(-1)} .$$

Our goal is a geometric classification of all supercommutative subalgebras of D.

Let us review what we did for the case of ordinary differential operators. We consider a commutative algebra B of ordinary differential operators. With a suitable coordinate change, we can assume without loss of generality that B has a *monic* operator, that is, an operator whose leading coefficient is 1. We say B_1 and B_2 are *isomorphic* if there is a function $f \in k[[x]]$ such that

$$B_1 = f^{-1} \cdot B_2 \cdot f .$$

The key point of the theory, due to I. Schur, is that for every such B, we can always find a function $f \in k[[x]]$ and a monic pseudodifferential operator

$$S = 1 + s_1 \left(\frac{d}{dx}\right)^{-1} + s_2 \left(\frac{d}{dx}\right)^{-2} + s_3 \left(\frac{d}{dx}\right)^{-3} + \cdots \tag{0.4}$$

of order 0, such that

$$A = S^{-1} \cdot f^{-1} \cdot B \cdot f \cdot S \subset k((\partial^{-1}))\,,$$

where $\partial = \frac{d}{dx}$. Thus all the information of a commutative algebra B of ordinary differential operator is encoded in the pair (A, S) consisting of a subalgebra

$$A \subset k((\partial^{-1}))$$

and an operator S of (0.4).

In order to obtain geometric information of the pair (A, S), we need yet another machinery involving the infinite-dimensional Grassmannian. Let us consider the quotient space E/Ex of E by the left maximal ideal Ex generated by x. The quotient space has a natural structure of a field. Actually, we can identify

$$E/Ex = k((\partial^{-1}))\,.$$

The set D of differential operators becomes $k[\partial]$ by the natural projection

$$\rho : E \longrightarrow E/Ex\,.$$

The Grassmannian we need, $G(0)$, is the set of all vector subspaces W of E/Ex which are commensurable with $\rho(D) = k[\partial]$. Since E acts on E/Ex from the left, E acts on the Grassmannian, too. Then we use the one-to-one correspondence between the big cell of the Grassmannian and the set of operators (0.4):

$$S \longmapsto S^{-1}\rho(D) = W \in G(0)\,.$$

Because of this correspondence, our pair (A, S) becomes a pair (A, W), which we called a *Schur pair* in **[M3]**. Then the relation $B \cdot D \subset D$ gives

$$(S^{-1} \cdot f^{-1} \cdot B \cdot f \cdot S) \cdot S^{-1}D \subset S^{-1}D\,.$$

Applying the projection ρ to the above, we obtain

$$A \cdot W \subset W\,.$$

At this point, it is easy to see how we construct from a Schur pair the geometric quintet we discussed in the beginning: the algebra A defines an affine curve $C \setminus \{p\}$, the inclusion $A \subset k((\partial^{-1}))$ gives a local covering π, the A-module W gives a torsion-free sheaf $\mathcal{F}$, and the inclusion $W \subset k((\partial^{-1}))$ determines a local isomorphism ϕ.

We want to follow the same path in the supersymmetric case. This time, we use an operator

$$S = 1 + s_1\delta^{-1} + s_2\delta^{-2} + s_3\delta^{-3} + \cdots\,. \tag{0.5}$$

We can show, after a suitable coordinate change and conjugation by a function $f \in k[[x]]$, that there exists an operator S of (0.5) for every supercommutative algebra B of super differential operators, such that

$$S^{-1} \cdot B \cdot S \subset k((\partial^{-1})) \oplus k((\partial^{-1}))\left(\xi\frac{\partial}{\partial\xi} - \frac{\partial}{\partial\xi}\xi\right) \oplus k((\partial^{-1}))\xi \oplus k((\partial^{-1}))\frac{\partial}{\partial\xi}\,. \tag{0.6}$$

We also have a natural isomorphism

$$(0.7)\qquad E/Ex \cong k((\partial^{-1})) \oplus k((\partial^{-1}))\left(\xi\frac{\partial}{\partial\xi} - \frac{\partial}{\partial\xi}\xi\right) \oplus k((\partial^{-1}))\xi \oplus k((\partial^{-1}))\frac{\partial}{\partial\xi}\,.$$

Thus we need to construct a suitable theory of Grassmannian based on the vector space appearing in the above. It has to be emphasized that the Grassmannian we obtain in this paper is not the super Grassmannian of [M4]. It is a much larger object, on which one can represent all the known supersymmetric KP systems as a simple system of vector fields.

Using our *noncommutative* Grassmannian, a new type of Schur pairs is introduced, and their geometric counterpart is studied. With these preparations, we can state our main theorem: *a supercommutative algebra of super differential operators is in one-to-one correspondence with a geometric quintet $(C, p, \pi, \mathcal{F}, \phi)$ consisting of a $\mathbb{Z}_2$-graded variety C of reduced dimension 1, a divisor $p \subset C$, a local covering π of C near p, an $\mathcal{O}_C$-module sheaf $\mathcal{F}$ satisfying the condition described below, and a local isomorphism ϕ.*

The sheaf $\mathcal{F}$ satisfies that there is a semi-stable torsion-free sheaf $\mathcal{F}_0$ on the reduced variety C_{red} satisfying the same cohomology vanishing of (0.1), such that

$$\mathcal{F} = \mathcal{F}_0 \otimes gl(1|1)\,,$$

where $gl(1|1)$ is the algebra of superlinear transformations of $\mathbb{A}_k^{1|1}$.

This paper is organized as follows. In Section 1, we give the definition of super pseudo-differential operators, and give the proof of the fact (0.6). The Grassmannian based on the space (0.7) is studied in Section 2. In Section 3, we establish the equivalence between the algebraic data of Schur pairs and the geometric quintets consisting of graded varieties and vector bundles on them. The main classification theorem is proved in Section 4. We also give a unified, geometric picture of the various super KP systems in this section.

ACKNOWLEDGEMENT: Part of the current paper has been worked out during my stay at the Mathematical Sciences Research Institute, Berkeley, as a member participating in the program of Strings in Mathematics and Physics. I wish to express my gratitude to MSRI for letting me use an office space and other facilities. A lot of conversations with Albert S. Schwarz were particularly useful for this work, most of which took place on the road from Davis to MSRI. I thank him for all of the interesting conversations.

1. $\mathbb{Z}_2$-commuting super differential operators.

Throughout this paper, k denotes an arbitrary field of characteristic zero. Whenever we say *graded* in this paper, it means $\mathbb{Z}_2$-graded, and it is equivalent with *supersymmetric* in physics language. The graded commutator, or the *supercommutator*, acting on a graded k-algebra $X = X_0 \oplus X_1$ is defined by

$$\begin{cases} [x,y] = xy - yx = -\,[y,x] & \text{if } x \in X_0 \\ [x,y] = xy + yx = [y,x] & \text{if } x \in X_1 \text{ and } y \in X_1\,. \end{cases}$$

We extend this definition k-linearly to the general element of X. We say $x \in X$ and $y \in X$ are $\mathbb{Z}_2$-commuting, or *supercommuting*, if $[x,y] = 0$. A subalgebra $Y \subset X$ is said to be a graded subalgebra, or a *super* subalgebra, if

$$Y = (Y \cap X_0) \oplus (Y \cap X_1)\,.$$

A graded subalgebra Y of X is $\mathbf{Z}_2$-commutative, or *supercommutative*, if $[y, y'] = 0$ for all $y, y' \in Y$.

In this section, we first define the algebra E of formal super pseudodifferential operators following Manin-Radul [**MaR**]. The set D of super differential operators forms a subalgebra of E. We then study $\mathbf{Z}_2$-commutative subalgebras of D, and determine their algebraic structure.

Our super pseudodifferential operators have coefficients in the $\mathbf{Z}_2$-commutative algebra

$$R = k[[x, \xi]] = k[[x]] \oplus k[[x]]\xi = R_0 \oplus R_1$$

of formal power series in an even variable x and an odd variable ξ. These variables satisfy $x \cdot \xi = \xi \cdot x$ and $\xi^2 = 0$. An element of R_0 (resp. R_1) is called a homogeneous element of degree 0 (resp. degree 1). We define the graded derivation operator δ by

$$\delta = \frac{\partial}{\partial \xi} + \xi \frac{\partial}{\partial x} .$$

It acts on the ring R by the graded Leibniz rule:

$$\delta(ab) = \delta(a) \cdot b + (-1)^{\tilde{a}} a\delta(b) ,$$

where a is a homogeneous element of R of $\mathbf{Z}_2$-degree $\tilde{a}$, and b is an arbitrary element of R. We have $\delta^2 = \frac{\partial}{\partial x}$. We call an expression

$$P = \sum_{m=0}^{\infty} a_m \delta^{n-m} \tag{1.1}$$

a *super pseudodifferential operator* with coefficients in R if $a_m \in R$. The *order* of P is defined to be n only when $0 \neq a_0 \in R_0$. The operator P of (1.1) is said to be *monic* if $a_0 = 1$, and *normalized* if $a_0 = 1$ and $a_1 = 0$. The set of all super pseudodifferential operators with coefficients in R is denoted by E. For an arbitrary integer ν and a nonnegative integer i, we define the graded binomial coefficients following [**MaR**] by

$$\begin{bmatrix} \nu \\ i \end{bmatrix} = \begin{cases} 0 & \text{if } 0 \leq \nu < i \text{ or } (\nu, i) \equiv (0, 1) \bmod 2 \\ \binom{[\frac{\nu}{2}]}{[\frac{i}{2}]} & \text{otherwise,} \end{cases}$$

where $[\alpha]$ denotes the largest integer not greater than α. The set E of super pseudo-differential operators has a graded algebra structure introduced by the generalized graded Leibniz rule:

$$\delta^{\nu} \cdot f = \sum_{i=0}^{\infty} (-1)^{\tilde{f} \cdot (\nu - i)} \begin{bmatrix} \nu \\ i \end{bmatrix} f^{[i]} \delta^{\nu - i} ,$$

where ν is an arbitrary integer, f is a homogeneous element of R of degree $\tilde{f}$, and $f^{[i]} = \delta^i(f)$.

Let $E^{(n)}$ denote the set of all super pseudodifferential operators of the form of (1.1). We have a natural filtration

$$\cdots \supset E^{(n+1)} \supset E^{(n)} \supset E^{(n-1)} \supset \cdots \tag{1.2}$$

of E which satisfies

$$\bigcup_{n\in\mathbf{Z}} E^{(n)} = E \quad \text{and} \quad \bigcap_{n\in\mathbf{Z}} E^{(n)} = \{0\} .$$

Thus E is a complete topological space. Let us define

$$(1.3) \qquad \begin{aligned} E_0 &= \{\sum_\nu f_\nu \delta^\nu \mid \tilde{f}_{2\nu} = 0 \text{ and } \tilde{f}_{2\nu+1} = 1\} , \\ E_1 &= \{\sum_\nu f_\nu \delta^\nu \mid \tilde{f}_{2\nu} = 1 \text{ and } \tilde{f}_{2\nu+1} = 0\} . \end{aligned}$$

Then $E = E_0 \oplus E_1$. An element of E_0 (resp. E_1) is called a *homogeneous-even* (resp. *homogenous-odd*) operator.

Symbolically, we can write $E = R((\delta^{-1}))$, where $k((x))$ is the standard notation for the field of fractions of the power series ring $k[[x]]$. Since $\frac{\partial}{\partial x} = \delta^2$ and $\frac{\partial}{\partial \xi} = \delta - \xi\delta^2$, we have

$$\begin{aligned} E = R((\delta^{-1})) &= R((\delta^{-2})) \oplus R((\delta^{-2}))(\delta - \xi\delta^2) \\ &= R((\partial^{-1})) \oplus R((\partial^{-1}))\frac{\partial}{\partial\xi} , \end{aligned}$$

where $\partial = \frac{\partial}{\partial x}$.

A rich geometric information is hidden in the set D of super differential operators. We call an element $P = \sum_\nu a_\nu \delta^\nu \in E$ a *super differential operator* if $a_\nu = 0$ for all negative ν.

Our goal of this paper is to give a geometric description of $\mathbf{Z}_2$-commutative graded subrings of D. Let us take an arbitrary odd parameter λ satisfying $\lambda^2 = 0$, and consider $\lambda \cdot D$. Certainly, this ring is a $\mathbf{Z}_2$-commutative ring of super differential operators, but it is not interesting at all. In order to avoid this trivial situation, we make the following assumption:

ASSUMPTION 1.4. *All the $\mathbf{Z}_2$-commutative subrings $B \subset D$ we consider in this paper satisfy the following conditions:*

(1) B is a $\mathbf{Z}_2$-commutative graded k-subalgebra of D containing k.

(2) B has a normalized homogeneous-even operator of order $2n$ for some $n > 0$.

Condition (2) is not a technical restriction. In fact, as we see in the below, we can always transform a monic operator into a normalized one by a simple coordinate change, and the monicness condition itself is not a strong restriction as explained in [M3]. For a subalgebra B of (1.4), we say that B is *nontrivial* if $B_1 = B \cap E_1 \neq 0$. Note that $B_0 = B \cap E_0 \neq k$ because of the above (2).

LEMMA 1.5. *Every monic, homogeneous-even super pseudodifferential operator P of order $2n$ can be transformed into a normalized operator by a suitable coordinate transformation.*

PROOF: Let

$$P = \delta^{2n} + a_1\delta^{2n-1} + a_2\delta^{2n-2} + \cdots .$$

Since $a_1 \in R_1$, it is of the form

$$a_1 = c\xi$$

with some $c \in k$. Therefore, the operator has the following form:

$$P = \partial^n + c\xi\frac{\partial}{\partial\xi}\partial^{n-1} + a_2\partial^{n-1} + \cdots .$$

Now define a coordinate transformation by

$$\begin{cases} x = y \\ \xi = e^{cy/n}\eta \, . \end{cases}$$

Then we have

$$\begin{aligned} \frac{\partial}{\partial y} &= \frac{\partial x}{\partial y}\frac{\partial}{\partial x} + \frac{\partial \xi}{\partial y}\frac{\partial}{\partial \xi} \\ &= \frac{\partial}{\partial x} + \frac{c}{n}\xi\frac{\partial}{\partial \xi} \, . \end{aligned}$$

Since

$$\left(\frac{\partial}{\partial y}\right)^n = \partial^n + c\xi\frac{\partial}{\partial \xi}\partial^{n-1} + \cdots \, ,$$

P has the desired normalized form in the new coordinate system. This completes the proof.

REMARK: We remark here that the above lemma is true for any odd element a_1 in a more general context. In our paper, however, we restrict ourselves to the consideration of operators defined over k.

Let $B \subset D$ be a subalgebra satisfying (1.4). Then it has a normalized homogeneous-even element P of order $2n$:

$$P = \delta^{2n} + 0 \cdot \delta^{2n-1} + a_2\delta^{2n-2} + \cdots \, .$$

It is easy to see that if we define $f = exp\big(- \frac{1}{n} \int a_2 dx \big)$, then we have

$$f^{-1} \cdot P \cdot f = \delta^{2n} + 0 \cdot \delta^{2n-1} + 0 \cdot \delta^{2n-2} + a_3\delta^{2n-3} + \cdots \, .$$

By a simple computation, one can obtain

LEMMA 1.6. *For a homogeneous-even super pseudodifferential operator* $Q = f^{-1} \cdot P \cdot f$ *of the form*

$$Q = \delta^{2n} + 0 \cdot \delta^{2n-1} + 0 \cdot \delta^{2n-2} + a_3\delta^{2n-3} + \cdots \, ,$$

there is a monic homogeneous-even operator

$$S = 1 + s_1\delta^{-1} + s_2\delta^{-2} + \cdots$$

of order 0 such that

$$S^{-1} \cdot Q \cdot S = \delta^{2n} \, .$$

The proof is given by an easy modification of [**M2**, Proposition 2.2]. Two $\mathbf{Z}_2$-commutative algebras B and B' of (1.4) are said to be *isomorphic* if there is an even element $f \in R_0$ such that

$$B' = f^{-1} \cdot B \cdot f \, .$$

Let us define

$$A = S^{-1} \cdot f^{-1} \cdot B \cdot f \cdot S \, .$$

Since every element of B commutes with P, everything in A commutes with δ^{2n}, and hence with $\frac{\partial}{\partial x}$. Therefore, A is a $\mathbf{Z}_2$-commutative subalgebra of

$$k((\partial^{-1})) \oplus k((\partial^{-1}))\left(\xi\frac{\partial}{\partial\xi} - \frac{\partial}{\partial\xi}\xi\right) \oplus k((\partial^{-1}))\xi \oplus k((\partial^{-1}))\frac{\partial}{\partial\xi}\,.$$

Let $A = A_0 \oplus A_1$, where A_0 (resp. A_1) is the homogenous-even (resp. homogenous-odd) part of A. Then we have

$$\begin{cases} A_0 \subset k((\partial^{-1})) \oplus k((\partial^{-1}))\left(\xi\frac{\partial}{\partial\xi} - \frac{\partial}{\partial\xi}\xi\right) \\ A_1 \subset k((\partial^{-1}))\xi \oplus k((\partial^{-1}))\frac{\partial}{\partial\xi}\,. \end{cases}$$

Now let us assume that our algebra B is nontrivial. Then $A_0 \neq k$ and $A_1 \neq 0$. Since $\xi\frac{\partial}{\partial\xi}$ and $\frac{\partial}{\partial\xi}\xi$ do not $\mathbf{Z}_2$-commute with ξ nor $\frac{\partial}{\partial\xi}$, A_0 has no element proportional to $\left(\xi\frac{\partial}{\partial\xi} - \frac{\partial}{\partial\xi}\xi\right)$. Similarly, since ξ and $\frac{\partial}{\partial\xi}$ do not $\mathbf{Z}_2$-commute one another, A_1 should be contained either in $k((\partial^{-1}))\xi$ or in $k((\partial^{-1}))\frac{\partial}{\partial\xi}$. Thus we have established the following:

THEOREM 1.7. *Let B be a nontrivial $\mathbf{Z}_2$-commutative graded subalgebra of D satisfying (1.4). Then there exist an invertible even element $f \in R_0$ and a homogenous-even operator*

$$S = 1 + s_1\delta^{-1} + s_2\delta^{-2} + \cdots$$

such that we have either

$$A = S^{-1} \cdot f^{-1} \cdot B \cdot f \cdot S \subset k((\partial^{-1})) \oplus k((\partial^{-1}))\xi\,,$$

or

$$A = S^{-1} \cdot f^{-1} \cdot B \cdot f \cdot S \subset k((\partial^{-1})) \oplus k((\partial^{-1}))\frac{\partial}{\partial\xi}\,.$$

With this theorem, we have determined the algebraic structure of a $\mathbf{Z}_2$-commutative algebras of super differential operators. For every B, we define its *rank* by

$$\text{rank } B = \text{rank } A = \frac{1}{2}\, G.C.D.\{\text{ord } a_0 \mid a_0 \in A_0\}\,. \tag{1.8}$$

The rank of B is always a positive divisor of n of (1.4).

2. The noncommutative Grassmannian.

In order to extract the geometric information of a $\mathbf{Z}_2$-commutative algebra of super differential operators, we need an intermediate step involving an infinite-dimensional Grassmannian. For the usual commuting ordinary differential operators, we used in [**M3**] the Grassmannian introduced by Sato [**Sa**], and established the complete geometric classification of commutative algebras of ordinary differential operators in terms of the Krichever functor. We then introduced the most natural supersymmetric generalization of the Grassmannian and the Krichever functor in [**M4**] and [**MR**]. The super Grassmannian was used effectively to establish the characterization of Jacobian varieties of the algebraic supercurves in [**M4**].

However, it turns out that the supersymmetric machinery of [**M4**] and [**MR**] is not suitable for our purpose of this paper. Of course one could extend the supersymmetric theory further so that it could provide a geometric framework for classifying our algebra B, but then such a theory would be rather ugly from the point of view of fanctoriality. Instead of going into such a direction, we present in this section a different idea of using a *noncommutative* Grassmannian.

Let us start with recalling the definition of the algebra E of super pseudodifferential operators:

$$E = (k[[x]] \oplus k[[x]]\xi)((\partial^{-1})) \oplus (k[[x]] \oplus k[[x]]\xi)((\partial^{-1}))\frac{\partial}{\partial\xi} .$$

The algebra generated by the four elements ξ, $\frac{\partial}{\partial\xi}$, $\xi\frac{\partial}{\partial\xi}$, and $\frac{\partial}{\partial\xi}\xi$ over k is denoted by $gl(1|1)$, which is the algebra of *superlinear* transformations of the $1|1$-dimensional affine superspace $\mathbb{A}_k^{1|1}$. One can introduce a matrix representation of $gl(1|1)$ by

$$\xi = \begin{bmatrix} 0 & 1 \\ 0 & 0 \end{bmatrix}, \quad \frac{\partial}{\partial\xi} = \begin{bmatrix} 0 & 0 \\ 1 & 0 \end{bmatrix}, \quad \xi\frac{\partial}{\partial\xi} = \begin{bmatrix} 1 & 0 \\ 0 & 0 \end{bmatrix}, \quad \frac{\partial}{\partial\xi}\xi = \begin{bmatrix} 0 & 0 \\ 0 & 1 \end{bmatrix} .$$

Using this algebra, we can give yet another presentation of E:

$$E = R_0((\partial^{-1})) \otimes_k gl(1|1) ,$$

where $R_0 = k[[x]]$. In this expression, we have

$$D = R_0[\partial] \otimes gl(1|1) \quad \text{and} \quad E^{(-1)} = R_0[[\partial^{-1}]]\partial^{-1} \otimes gl(1|1) .$$

There is a natural left R-module direct sum decomposition

$$E = D \oplus E^{(-1)} . \tag{2.1}$$

Let

$$\rho : E \longrightarrow E/Ex = V$$

be the natural projection, where Ex is the left ideal of E generated by x. We have a natural identification

$$\begin{aligned} V = E/Ex &= k((\partial^{-1})) \oplus k((\partial^{-1}))\left(\xi\frac{\partial}{\partial\xi} - \frac{\partial}{\partial\xi}\xi\right) \oplus k((\partial^{-1}))\xi \oplus k((\partial^{-1}))\frac{\partial}{\partial\xi} \\ &= k((\partial^{-1})) \otimes gl(1|1) . \end{aligned}$$

The direct sum decomposition (2.1) of E descends to V:

$$\begin{aligned} V &= V_+ \oplus V^{(-1)} \\ &= \rho(D) \oplus \rho(E^{(-1)}) \\ &= \left(k[\partial] \otimes gl(1|1)\right) \oplus \left(k[[\partial^{-1}]]\partial^{-1} \otimes gl(1|1)\right) . \end{aligned}$$

The Grassmannian we use in this paper is the following:

DEFINITION 2.2. *The noncommutative Grassmannian of index μ, which is denoted by $G(\mu)$, is the set of all right $gl(1|1)$ submodules W of V such that the natural projection*

$$\gamma_W : W \longrightarrow V/V^{(-1)}$$

satisfies that

$$\mathrm{rank}_{gl(1|1)} \mathrm{Ker}\, \gamma_W - \mathrm{rank}_{gl(1|1)} \mathrm{Coker}\, \gamma_W = \mu \, .$$

The big cell $G^+(0)$ of the noncommutative Grassmannian is the set of right $gl(1|1)$ submodules $W \subset V$ such that the projection γ_W is an isomorphism over $gl(1|1)$.

Our noncommutative Grassmannian is the set of all right $gl(1|1)$ submodules $W \subset V$ which differ from $\rho(D)$ by finite rank μ over $gl(1|1)$. Thus the noncommutative Grassmannian has the *base point* $\rho(D) \in G(0)$. We note here that E acts on V naturally from the left. Every element of E gives rise to a vector field on the Grassmannian. In order to see this, let us recall that the tangent space of $G(\mu)$ at W is given by

$$T_W G(\mu) = \mathrm{Hom}_{gl(1|1)}(W, V/W) \, ,$$

where $\mathrm{Hom}_{gl(1|1)}(W, V/W)$ denotes the set of all right $gl(1|1)$-homomorphisms of W into V/W. An operator $P \in E$ defines

$$\Phi_W(P) : W \hookrightarrow V \xrightarrow{P} V \to V/W \, ,$$

which is an element of $\mathrm{Hom}_{gl(1|1)}(W, V/W)$ because P acts on V from the left. Therefore, $P \in E$ determines a vector field on the noncommutative Grassmannian by

$$\Phi(P) : G(\mu) \ni W \longmapsto \Phi_W(P) \in \mathrm{Hom}_{gl(1|1)}(W, V/W) = T_W G(\mu) \, .$$

Following the argument of [**M3**, Section 7] almost literally, we can establish two theorems which play the key role in this paper:

THEOREM 2.3. *A super pseudodifferential operator $P \in E$ is a super differential operator (i.e. $P \in D$) if and only if P stabilizes the base point of the noncommutative Grassmannian: $\Phi_{\rho(D)}(P) = 0$.*

THEOREM 2.4. *There is a natural bijective correspondence between the group Γ_0 of monic homogeneous-even super pseudodifferential operators of order 0 and the big cell $G^+(0)$:*

$$\Gamma_0 \ni S \longmapsto S^{-1}\rho(D) \in G^+(0) \, .$$

With these preparation, let us go back to the final stage we reached at the end of Section 1. We have a $\mathbf{Z}_2$-commutative algebra $B \subset D$ and a prescribed element $P \in B$. Then Theorem 1.7 tells us that there are an even element $f \in R_0$ and an operator $S \in \Gamma_0$ such that

$$A = S^{-1} \cdot f^{-1} \cdot B \cdot f \cdot S$$

satisfies either $A \subset k((\partial^{-1})) \oplus k((\partial^{-1}))\frac{\partial}{\partial \xi}$ or $A \subset k((\partial^{-1})) \oplus k((\partial^{-1}))\xi$. Since $B \subset D$, we have $B \cdot D \subset D$. The fact that $f \cdot D = D$ implies that $(f^{-1} \cdot B \cdot f) \cdot D \subset D$. Therefore, we have

$$(S^{-1} \cdot f^{-1} \cdot B \cdot f \cdot S) \cdot S^{-1}D \subset S^{-1}D \, .$$

It means that the algebra A *stabilizes* the point $W = S^{-1}\rho(D)$ of the big cell $G^+(0)$ of the Grassmannian:

$$A \cdot W \subset W .$$

It should be noted that the data (A, W) are not uniquely determined by the algebra B. Actually, we transformed $P \in B$ into ∂^n in Section 1, but that is not necessary in order to make A a subset of V. In fact, every operator of E commutes with ∂ if it commutes with any nonconstant element of $k((\partial^{-1}))$. It motivates us to define the following:

DEFINITION 2.5. *A homogeneous-even zeroth order operator $T \in E$ is said to be admissible if it is invertible and satisfies that*

$$T^{-1} \cdot \partial \cdot T \in k((\partial^{-1})) .$$

It is easy to show that an admissible operator T has the following form:

$$T = e^{c_0 x} \left(\sum_{m=0}^{\infty} a_m(x)\partial^{-m} + \sum_{n=0}^{\infty} b_n(x)\xi \frac{\partial}{\partial \xi} \partial^{-n-1} \right) ,$$

where $a_n(x)$ and $b_n(x)$ are polynomials in x of degree less than or equal to n with constant coefficients, and c_0 is an arbitrary constant. In order for T to be invertible, we need $a_0 \neq 0$.

DEFINITION 2.6. *A Schur pair is a pair (A, W) consisting of a graded k-subalgebra $A = A_0 \oplus A_1$ of $k((\partial^{-1})) \oplus k((\partial^{-1}))\xi$ or $k((\partial^{-1})) \oplus k((\partial^{-1}))\frac{\partial}{\partial \xi}$, and a point W of the big cell of the noncommutative Grassmannian, satisfying that $A_0 \neq k$, $A_1 \neq 0$, and that*

$$A \cdot W \subset W .$$

Two Schur pairs (A, W) and (A', W') are said to be isomorphic if there is an admissible operator T such that

$$\begin{cases} A' = T^{-1} \cdot A \cdot T \\ W' = T^{-1} W \end{cases}$$

With these terminology, we can state

THEOREM 2.7. *There is a bijective correspondence between the set of isomorphism classes of the $\mathbf{Z}_2$-commutative algebras of super differential operators of (1.4), and the set of isomorphism classes of Schur pairs of (2.6).*

PROOF:Let B_1 be the algebra of (1.4), and choose $f_1 \in R_0$ and $S_1 \in \Gamma_0$ as in Theorem 1.7. Then the corresponding Schur pair is given by

$$(A_1, W_1) = (S_1^{-1} \cdot f_1^{-1} \cdot B_1 \cdot f_1 \cdot S_1,\ S_1^{-1}\rho(D)) .$$

Take another algebra B_2 isomorphic to B_1. Then it gives a different Schur pair

$$(A_2, W_2) = (S_2^{-1} \cdot f_2^{-1} \cdot B_2 \cdot f_2 \cdot S_2,\ S_2^{-1}\rho(D)) .$$

Since $B_2 = f^{-1} \cdot B_1 \cdot f$ for some $f \in R_0$, we have

$$A_2 = S_2^{-1} \cdot f_2^{-1} \cdot f^{-1} \cdot f_1 \cdot S_1 \cdot A_1 \cdot S_1^{-1} \cdot f_1^{-1} \cdot f \cdot f_2 \cdot S_2 .$$

Therefore, $T = S_1^{-1} \cdot f_1^{-1} \cdot f \cdot f_2 \cdot S_2$ is an admissible operator. Note that we have $W_2 = T^{-1} \cdot W_1$. Thus (A_1, W_1) and (A_2, W_2) are isomorphic.

Now let us show the converse. So take a Schur pair (A, W). There is an $S \in \Gamma_0$ corresponding to W such that $W = S^{-1}\rho(D)$. Since $A \cdot W \subset W$, we have

$$(S \cdot A \cdot S^{-1}) \cdot \rho(D) \subset \rho(D) .$$

But this means that the algebra

$$B = S \cdot A \cdot S^{-1}$$

stabilizes the base point $\rho(D)$. Therefore, B consists of super differential operators. It is easy to see that B satisfies (1.4). Let us choose an admissible operator T, and consider the Schur pair $(T^{-1} \cdot A \cdot T,\ T^{-1} \cdot W)$ isomorphic to the original one. The point $T^{-1} \cdot W$ of the big cell corresponds to an operator

$$a_0^{-1} \cdot S \cdot T \in \Gamma_0 ,$$

where $a_0 \in R_0$ is the leading term of T. This pair gives rise to an algebra

$$(a_0^{-1} \cdot S \cdot T) \cdot T^{-1} \cdot A \cdot T \cdot (T^{-1} \cdot S^{-1} \cdot a_0) = a_0^{-1} \cdot B \cdot a_0 ,$$

which is isomorphic to B. This completes the proof.

This theorem tells us that the Schur pair (A, W) possesses all the information that a $\mathbf{Z}_2$-commutative algebra of super differential operators has. In order to extract its geometric information, we need a suitable generalization of the Krichever functor of [**M3**].

3. The noncommutative version of the Krichever correspondence.

In the usual case of ordinary differential operators, a commutative algebra of ordinary differential operators corresponds to a set of geometric data consisting of an algebraic curve and a semi-stable vector bundle on it. However, in our case of super differential operators, a $\mathbf{Z}_2$-commutative graded subalgebra of D *does not* correspond to a set of data consisting of an algebraic supercurve and a super vector bundle on it. The geometric data appearing naturally in our case are rather different from what we have dealt with in [**MR**], and this is the topic of this section. We have established in the last section that an algebra B of (1.4) is in one-to-one correspondence with a Schur pair (A, W). In this section, firstly, we define a category of Schur pairs incorporating the notion of isomorphism in a more natural way. We also define a geometric category of algebraic curves and vector bundles on them in a suitable way to our current situation, and then establish antiequivalence of these categories.

In order to define a category of Schur pairs, we have to relax the condition of the Schur pairs given in Definition 2.6. We call (A, W) a Schur pair of *rank* r and *index* μ, if (i) A is a $\mathbf{Z}_2$-commutative subalgebra of $V = k((\partial^{-1})) \otimes gl(1|1)$ such that $A_0 = A \cap k((\partial^{-1})) \supset k$, $A_0 \neq k$, $A_1 = A \cap k((\partial^{-1}))\xi \oplus k((\partial^{-1}))\frac{\partial}{\partial\xi} \neq 0$, and

$$r = \text{rank } A = \frac{1}{2}\, G.C.D.\{\text{ord } a_0 \mid a_0 \in A_0\} ;$$

and (ii) $W \in G(\mu)$. (Note that we have defined ord ∂ = ord $\delta^2 = 2$.)

The *Category of Schur pairs*, which we denote by $\mathcal{S}$, has a Schur pair of an arbitrary positive rank and an arbitrary index as its object. A morphism between two Schur pairs (A, W) and (A', W') is a set (α, ι) of *twisted inclusions*

$$\text{(3.1)} \qquad \begin{cases} \alpha : T^{-1} \cdot A \cdot T \hookrightarrow A' \\ \iota : T^{-1} W \hookrightarrow W' . \end{cases}$$

Note here that we are not requiring that W is a point of the big cell any more. We also note that A is a subalgebra of either $k((\partial^{-1})) \oplus k((\partial^{-1}))\xi$ or $k((\partial^{-1})) \oplus k((\partial^{-1}))\frac{\partial}{\partial \xi}$ as in Section 1. If $A \subset k((\partial^{-1})) \oplus k((\partial^{-1}))\xi$ and $A' \subset k((\partial^{-1})) \oplus k((\partial^{-1}))\frac{\partial}{\partial \xi}$ (or the other way around), then there is no morphism between the Schur pairs (A, W) and (A', W').

Next, let us define the category of the geometric objects corresponding to the Schur pairs. The category $\mathcal{Q}$ consists of a *quintet* $(C, p, \pi, \mathcal{F}, \phi)$ of an arbitrary positive *rank* r and an arbitrary *index* $\mu_0|\mu_1$, where

(3.2) $C = (C_{\text{red}}, \mathcal{O}_C)$ is a graded algebraic variety of even-part dimension 1 defined over k such that C_{red} is a reduced irreducible complete algebraic curve over k, and the structure sheaf is given by $\mathcal{O}_C = \mathcal{O}_{C_{\text{red}}} \oplus \mathcal{N}$, where $\mathcal{N}$ is a nonzero torsion-free sheaf of $\mathcal{O}_{C_{\text{red}}}$-modules. We require that $\mathcal{O}_C$ is a $\mathbf{Z}_2$-commutative subalgebra of $\mathcal{O}_{C_{\text{red}}} \otimes gl(1|1)$. It follows from this requirement that $\mathcal{N}$ is a nilpotent algebra: $\mathcal{N}^2 = 0$.

(3.3) $p \subset C$ is a divisor of C such that its reduced point p_{red} is a smooth k-rational point of C_{red}.

(3.4) $\pi : U_0 \to U_{\text{red},p_{\text{red}}}$ is a morphism of formal schemes, where $U_0 \cong \operatorname{Spec} k[[\partial^{-1}]]$ is the formal completion of the affine line $\mathcal{A}_k^1$ at the origin, and $U_{\text{red},p_{\text{red}}}$ is the formal completion of the algebraic curve C_{red} at the point p_{red}. We require that π is an r-sheeted covering ramified at p_{red}.

(3.5) $\mathcal{F}$ is a sheaf of $\mathcal{O}_C$-modules defined on C. As an $\mathcal{O}_{C_{\text{red}}}$-module, $\mathcal{F}$ is torsion-free. It also has an $\mathcal{O}_{C_{\text{red}}} \otimes gl(1|1)$-module structure with generic rank r, and satisfies that

$$\operatorname{rank}_{gl(1|1)} H^0(C_{\text{red}}, \mathcal{F}) - \operatorname{rank}_{gl(1|1)} H^1(C_{\text{red}}, \mathcal{F}) = \mu .$$

(3.6) ϕ is a sheaf isomorphism

$$\phi : \mathcal{F} \otimes_{\mathcal{O}_{C_{\text{red}}}} \mathcal{O}_{U_{\text{red},p'_{\text{red}}}} \xrightarrow{\sim} \pi_* \mathcal{O}_{U_0}(-1) \otimes gl(1|1) .$$

Two quintets $(C, p, \pi_1, \mathcal{F}, \phi_1)$ and $(C, p, \pi_2, \mathcal{F}, \phi_2)$ are identified if the diagram

$$\begin{array}{ccc} H^0(U_{\text{red},p_{\text{red}}}, \mathcal{F} \otimes \mathcal{O}_{U_{\text{red},p_{\text{red}}}}) & \xrightarrow[\sim]{\phi_1} & H^0(U_{\text{red},p_{\text{red}}}, \pi_{1*}\mathcal{O}_{U_0}(-1) \otimes gl(1|1)) \\ {\scriptstyle \phi_2}\downarrow{\scriptstyle \wr} & & \downarrow{\scriptstyle \wr} \\ H^0(U_{\text{red},p_{\text{red}}}, \pi_{2*}\mathcal{O}_{U_0}(-1) \otimes gl(1|1)) & \xrightarrow{\sim} & H^0(U_0, \mathcal{O}_{U_0}(-1) \otimes gl(1|1)) . \end{array}$$

commutes.

A morphism

$$(\beta, \psi) : (C', p', \mathcal{F}', \pi', \phi') \longrightarrow (C, p, \mathcal{F}, \pi, \phi)$$

of two quintets consists of a morphism $\beta : C' \to C$ of graded algebraic varieties and a homomorphism $\psi : \mathcal{F} \to \beta_* \mathcal{F}'$ of sheaves on C satisfying the following conditions:

(3.7) The divisor p' is the pull-back of p: $p' = \beta^{-1}(p)$.

(3.8) There exists a formal scheme isomorphism $h : U_0 \xrightarrow{\sim} U_0$ and a nonzero constant $c \in k^*$ such that

$$\begin{array}{ccc} U_0 & \xrightarrow{ch} & U_0 \\ \pi' \downarrow & & \downarrow \pi \\ U_{\text{red},p'_{\text{red}}} & \xrightarrow{\widehat{\beta}} & U_{\text{red},p_{\text{red}}} , \end{array}$$

where $\widehat{\beta}$ is the morphism of formal schemes determined by β.

(3.9) There is an $\mathcal{O}_{U_0}$-module isomorphism $\zeta : \mathcal{O}_{U_0}(-1) \xrightarrow{\sim} (ch)_* \mathcal{O}_{U_0}(-1)$ such that

$$\begin{array}{ccccc} \mathcal{F} \otimes \mathcal{O}_{U_{\text{red},p_{\text{red}}}} & \xrightarrow{\widehat{\psi}} & \widehat{\beta}_* \mathcal{F}' \otimes \mathcal{O}_{U'_{\text{red},p'_{\text{red}}}} & & \\ \phi \downarrow & & \downarrow \widehat{\beta}_*(\phi') & & \\ \pi_* \mathcal{O}_{U_0}(-1) \otimes gl(1|1) & \xrightarrow{\pi_*(\zeta)} & \pi_*(ch)_* \mathcal{O}_{U_0}(-1) \otimes gl(1|1) & = & \widehat{\beta}_* \pi'_* \mathcal{O}_{U_0}(-1) \otimes gl(1|1) , \end{array}$$

where $\widehat{\psi}$ is the homomorphism of sheaves on $U_{\text{red},p_{\text{red}}}$ defined by ψ.

Our graded variety is rather different from what is called the *supermanifold*. Actually, our variety C has no supermanifold structure, even locally, in general, because of the nilpotency of $\mathcal{N}$ of (3.2).

Now we can state the main theorem of this section:

THEOREM 3.10. *There is a fully-faithful contravariant functor, which we call the graded Krichever functor, between the categories $\mathcal{Q}$ and $\mathcal{S}$;*

$$\chi : \mathcal{Q} \xrightarrow{\sim} \mathcal{S} .$$

PROOF: First of all, we identify

$$U_0 = \text{Spec } k[[\partial^{-1}]] .$$

This identification together with the isomorphism of (3.6) makes

$$W = \phi(H^0(C_{\text{red}} \setminus \{p_{\text{red}}\}, \mathcal{F}))$$

a $gl(1|1)$ submodule of V. The condition of the Euler characteristic of $\mathcal{F}$ imposed in (3.5) dictates that W is a point of the noncommutative Grassmannian $G(\mu)$.

We define the algebra A by

$$A = \pi^*((H^0(C_{\text{red}} \setminus \{p_{\text{red}}\}, \mathcal{O}_C)) .$$

The definition of π given in (3.4) makes A a subalgebra of V. It is easy to see that (A, W) becomes a Schur pair of rank r and index μ.

Thus we have defined a map

$$\chi : (C, p, \pi, \mathcal{F}, \phi) \longmapsto (A, W) ,$$

which gives a functor of $\mathcal{Q}$ into $\mathcal{S}$. This is really a cohomology functor, and therefore its fanctoriality and the correspondence between the morphisms can be established using the cohomology theory along the line of [**M3**, Theorem 4.6].

In order to show that χ is fully-faithful, we have to give its inverse. The construction of C_{red} from A_0 is exactly the same as in [**M3**, Section 3]. The A_0-module A_1 gives the nilpotent sheaf $\mathcal{N}$, and the A-module W determines $\mathcal{F}$. The embedding of A_0 in the larger ring $k((\partial^{-1}))$ defines the local covering map π. Similarly, the inclusion $W \hookrightarrow V$ gives ϕ. Because of the fanctoriality, one can establish the antiequivalence of $\mathcal{Q}$ and $\mathcal{S}$. This completes the proof.

4. The main theorem.

We are almost ready to state the main classification theorem of this paper now. Before doing that, we need a couple of more notations.

We denote by $\mathcal{B}$ the set of isomorphism classes of $\mathbf{Z}_2$-commutative algebras B of (1.4). Let us recall that two algebras B and B' are said to be isomorphic if there is a function $f \in R_0$ such that $B' = f^{-1} \cdot B \cdot f$. An element of $\mathcal{B}$ corresponds bijectively to an isomorphism class of certain Schur pairs by Theorem 2.7. And by Theorem 3.10, an isomorphism class of Schur pairs corresponds, again bijectively, to an isomorphism class of quintets.

So let us denote by $\mathcal{M}^+(0)$ the set of isomorphism classes of quintets $(C,\, p,\, \pi,\, \mathcal{F},\, \phi)$ such that the sheaf $\mathcal{F}$ satisfies that

$$H^0(C_{\text{red}}, \mathcal{F}) = H^1(C_{\text{red}}, \mathcal{F}) = 0\,. \tag{4.1}$$

Because of (3.5), there is a sheaf $\mathcal{F}_0$ of torsion-free $\mathcal{O}_{C_{\text{red}}}$-modules of rank r such that

$$\mathcal{F} \cong \mathcal{F}_0 \otimes gl(1|1)\,.$$

The cohomology vanishing of (4.1) then implies that $\mathcal{F}_0$ is a semi-stable sheaf of degree $r(g-1)$, where g is the arithmetic genus of C_{red} [**M3**, Proposition 3.8].

By the graded Krichever functor χ, an element of $\mathcal{M}^+(0)$ corresponds to an isomorphism class of a Schur pair (A, W) with $W \in G^+(0)$. Note that the action $T^{-1}W$ of an admissible operator T on the Grassmannian stabilizes the index and the big cell of the Grassmannian.

Thus we have established the geometric classification of $\mathbf{Z}_2$-commutative algebras of super differential operators:

THEOREM 4.2. *There is a natural bijective correspondence between $\mathcal{B}$ and $\mathcal{M}^+(0)$:*

$$\mu : \mathcal{B} \xrightarrow{\sim} \mathcal{M}^+(0)\,.$$

In the rest of this section, we first study maximal elements of $\mathcal{B}$, and then study various super KP systems in terms of vector fields on the noncommutative Grassmannian.

An element of $\mathcal{B}$ is said to be *maximal* if it is a maximal element with respect to the twisted inclusion relation

$$B' \subset f^{-1} \cdot B \cdot f$$

for some $f \in R_0$. We call a Schur pair (A, W) *maximal* if A is a maximal $\mathbf{Z}_2$-commutative stabilizer of W, i.e.,

$$A = \{a \in V \mid a \cdot W \subset W\}\,,$$

where

$$V = k((\partial^{-1})) \oplus k((\partial^{-1}))\left(\xi\frac{\partial}{\partial\xi} - \frac{\partial}{\partial\xi}\xi\right) \oplus k((\partial^{-1}))\xi \oplus k((\partial^{-1}))\frac{\partial}{\partial\xi}\,.$$

If $B \in \mathcal{B}$ is maximal, then the Schur pair (A, W) corresponding to B is maximal, and vice versa.

The algebra $gl(1|1)$ has an involution

$$\sigma : gl(1|1) \xrightarrow{\sim} gl(1|1)$$

defined by interchanging ξ and $\frac{\partial}{\partial \xi}$. The involution σ induces involutions in E, in $\mathcal{B}$, and in the categories $\mathcal{Q}$ and $\mathcal{S}$. If B is maximal, so is $\sigma^{-1}B$. If B corresponds to (A, W) with $A \subset k((\partial^{-1})) \oplus k((\partial^{-1}))\xi$, then $\sigma^{-1}B$ corresponds to $(\sigma^{-1}A, W)$, and we have $\sigma^{-1}A \subset k((\partial^{-1})) \oplus k((\partial^{-1}))\frac{\partial}{\partial \xi}$ this time. Certainly A and $\sigma^{-1}A$ are isomorphic as a graded algebra. Therefore, B and $\sigma^{-1}B$ are also isomorphic, but it is not so obvious from the Manin-Radul description of the super differential operators.

Thus the maximal elements of $\mathcal{B}$ are in two-to-one correspondence with the points of the big cell of the noncommutative Grassmannian, but, of course, this correspondence is not surjective.

A maximal Schur pair satisfies an interesting property (cf. [**MR**, Theorem 1.3]):

PROPOSITION 4.3. *Let (A, W) be a maximal Schur pair, and $A = A_0 \oplus A_1$ the decomposition of A into its even part and odd part. Then A_1 is an A_0-module of rank one.*

PROOF: Let r be the rank of (A, W). Then for every large $n \in \mathbb{N}$, A_0 has an element whose leading term is ∂^{nr} [**M3**, Proposition 3.2]. Since σ stabilizes W, $(\sigma^{-1}A, W)$ gives the other maximal Schur pair having the same W. Certainly, we have $\sigma^{-1}A = A_0 \oplus \sigma^{-1}A_1$. Therefore, we can assume without loss of generality that $A_1 \subset k((\partial^{-1}))\xi$.

Take a nonzero element $a\xi \in A_1$, where $a = a(\partial) \in k((\partial^{-1}))$. In order to show that A_1 has rank one over A_0, it suffices to show that

$$\dim_k A_1 / a\xi A_0 < +\infty \, .$$

Since $a\frac{\partial}{\partial \xi} \in \sigma^{-1}A$, it stabilizes W. Thus $\left[a\frac{\partial}{\partial \xi}, A\right] \subset \sigma^{-1}A$, and hence $\left[a\frac{\partial}{\partial \xi}, A_1\right] \subset A_0$ because of the maximality of A. Note that we have

$$\left[\frac{\partial}{\partial \xi}, \xi\right] = 1 \, .$$

Applying the operation $\left[a\frac{\partial}{\partial \xi}, \cdot\right]$ to the inclusion relation $a\xi A_0 \subset A_1$, we obtain

$$a^2 A_0 \subset \left[a\frac{\partial}{\partial \xi}, A_1\right] \subset A_0 \, .$$

But since $A_0/a^2 A_0$ is finite over k, so is $A_1/a\xi A_0$. This completes the proof.

It follows from this proposition that the graded variety C of the quintet corresponding to a maximal Schur pair is an algebraic supercurve of dimension $1|1$, because its structure sheaf satisfies

$$\mathcal{O}_C = \mathcal{O}_{C_{\mathrm{red}}} \oplus \mathcal{N} \cong \wedge^\bullet(\mathcal{N})$$

with a torsion-free rank 1 sheaf $\mathcal{N}$ of $\mathcal{O}_{C_{\mathrm{red}}}$-modules. If further C_{red} is nonsingular, then C is nothing but a supermanifold of dimension $1|1$ in the sense of [**Ma**]. The sheaf $\mathcal{F}$ is then a super vector bundle of rank $r|r$. It is an abuse of terminology, but we can safely call $\mathcal{F}$

a super vector bundle over C even if C_{red} is singular, because $\mathcal{F}$ is indeed torsion-free over $\mathcal{O}_{C_{red}}$ in any case. With these terminology, we have the following:

THEOREM 4.4. *Every maximal element of $\mathcal{B}$ is in one-to-one correspondence with a quintet $(C, p, \pi, \mathcal{F}, \phi) \in \mathcal{M}^+(0)$ such that C is an algebraic supercurve of dimension $1|1$ over k, p is a $0|1$-dimensional divisor of C, and $\mathcal{F}$ is a super vector bundle over C of rank $r|r$.*

Finally, let us study relations of our theory with the various known super KP systems of [**KL**], [**MaR**], [**M4**], and [**R**]. As we have noted in Section 2, every element P of E defines a vector field $\Phi(P)$ on the noncommutative Grassmannian, and hence on the big cell $G^+(0)$. We can interpret the vector field in terms of a differential equation of a monic homogeneous-even super pseudodifferential operator $S \in \Gamma_0$ of order 0 by Theorem 2.4. In the notation of [**M2**], the equation is given by

$$\frac{\partial S}{\partial t_P} = -(S \cdot P \cdot S^{-1})_- \cdot S ,$$

where Q_- represents the $E^{(-1)}$-part of $Q \in E$ following the decomposition of (2.1), and t_P is the parameter corresponding to the vector field $\Phi(P)$. It is obvious from the equation that if P is in $E^{(-1)}$, then the equation can be integrated immediately:

$$S(t_P) = S \cdot e^{-t_P P} . \tag{4.5}$$

Therefore, the equation is interesting only when $P \in D$.

If we have a commutative set $K \subset D$ of super differential operators, then it induces a commutative system of vector fields on the Grassmannian, and a compatible system of nonlinear partial differential equations on $S \in \Gamma_0$. If we take K to be supercommutative, then the corresponding system is compatible in the supersymmetric sense.

The super KP system that Rabin [**R**] and I [**M4**] discovered independently is given by taking

$$K = k[\partial] \oplus k[\partial]\frac{\partial}{\partial \xi} \subset D ,$$

which is a system of maximally compatible nonlinear super partial differential equations on $S \in \Gamma_0$. Applying the involution σ, one obtains yet another compatible system:

$$\sigma^{-1} K = k[\partial] \oplus k[\partial]\xi .$$

The compatible system introduced by Manin and Radul [**MaR**] is the one which mixes these two systems. With the ingenious infinite sums, their system remains compatible.

It has been shown by Schwarz [**Sc2**] that if one takes

$$\begin{aligned} K &= k[\partial] \oplus k[\partial]\left(\xi\frac{\partial}{\partial\xi} - \frac{\partial}{\partial\xi}\xi\right) \oplus k[\partial]\xi \oplus k[\partial]\frac{\partial}{\partial\xi} \\ &= k[\partial] \otimes gl(1|1) \subset D , \end{aligned}$$

then it gives the Kac-van de Leur super KP system [**KL**], which is no longer compatible as a system of super partial differential equations. From the point of view of analysis, the Kac-van de Leur system is not natural because of the noncompatibility, but from the algebraic point of view, it is natural because of the universality. Namely, it generates every flow on the Grassmannian $G(0)$ that does not have a simple integral of (4.5).

Each of these systems defines a *deformation theory* of $\mathcal{B}$ and $\mathcal{Q}$. In particular, these deformations can be described in terms of geometric deformations of the graded algebraic variety C and the sheaf $\mathcal{F}$ on it. All of these deformations preserve the reduced curve C_{red}, but the graded structure $\mathcal{O}_C$ as well as the sheaf $\mathcal{F}$ are deformed. For a compatible deformations, we can talk about the orbit of the flows, which gives a deformation space. However, these deformation spaces do not have a simple geometric description, such as *Jacobian varieties*, in general. Our new noncommutative Grassmannian provides a systematic and unified picture for all the known supersymmetric KP systems.

But if one wants to study *geometry* of Jacobian varieties of supercurves, then the current picture we have established in this paper is not useful. For that purpose, one needs the super Grassmannian and the Jacobian flows of [M4]. On the other hand, now we can understand why the machinery of [M4] does not provide any clear picture for the Kac-van de Leur system. Actually, we can see only half of the possible deformations in terms of the super Grassmannian picture.

The deformations of an algebra B by these flows can be interpreted as *isospectral* deformations, because A_0 represents the spectral structure of B. It is easy to see that the Kac-van de Leur flows give all possible isospectral deformations of super differential operators in this sense.

References

[K] I. M. Krichever: Methods of algebraic geometry in the theory of nonlinear equations, Russ. Math. Surv. **32** (1977) 185–214.

[KL] V. G. Kac and J. W. van de Leur: Super boson-fermion correspondence of type B, Adv. Ser. in Math. Phys. **7**, World Scientific, 1989.

[Ma] Yu. I. Manin: Gauge field theory and complex geometry, Grundlehren der mathematischen Wissenschaften **289**, Springer-Verlag (1988) 297pps.

[MaR] Yu. I. Manin and A. O. Radul: A supersymmetric extension of the Kadomtsev-Petviashvili hierarchy, Commun. Math. Phys. **98** (1985) 65–77.

[M1] M. Mulase: Cohomological structure in soliton equations and jacobian varieties, J. Differential Geom. **19** (1984) 403–430.

[M2] M. Mulase: Solvability of the super KP equation and a generalization of the Birkhoff decomposition, Invent. Math. **92** (1988) 1–46.

[M3] M. Mulase: Category of vector bundles on algebraic curves and infinite dimensional Grassmannians, Intern. J. of Math. **1** (1990) 293–342.

[M4] M. Mulase: A new super KP system and a characterization of the Jacobians of arbitrary algebraic supercurves, J. Differential Geom. **34** (1991) 651–680.

[MR] M. Mulase and J. M. Rabin: Super Krichever functor, Intern. J. of Math. **2** (1991) 741–760.

[N1] A. Nakayashiki: Structure of Baker-Akhiezer modules of principally polarized Abelian varieties, commuting partial differential operators and associated integrable systems, Duke Math. J. **62** (1991)

[N2] A. Nakayashiki: Commuting partial differential operators and vector bundles over Abelian varieties, Amer. J. Math., to appear.

[R] J. M. Rabin: The geometry of the super KP flows, Commun. Math. Phys. **137** (1991) 533–552.

[Sa] M. Sato: Soliton equations as dynamical systems on an infinite dimensional Grassmann manifold, Kokyuroku, Res. Inst. Math. Sci., Kyoto Univ. **439** (1981) 30–46.

[Sc1] A. S. Schwarz: On solutions to the string equation, MSRI preprint 05429–91 (1991).

[Sc2] A. S. Schwarz: (in preparation).

[SW] G. B. Segal and G. Wilson: Loop groups and equations of KdV type, Publ. Math. I. H. E. S. **61** (1985) 5–65.

13

Relative Bounds for Fano Varieties of the Second Kind

Alan M. Nadel The Institute for Advanced Study, Princeton, New Jersey

Introduction

This article is concerned with bounding Fano varieties of the second kind. Recall that a *Fano variety* is a smooth projective variety with ample anticanonical class. A Fano variety is said to be of the *first kind* if its Picard number is one, and of the *second kind* if its Picard number is at least two. Fano varieties of the second kind have nontrivial extremal rays, which can be contracted in accordance with Mori's program, and are thus sometimes regarded as being less primitive than Fano varieties of the first kind.

There is a well-known conjecture asserting that there are only finitely many deformation types of Fano varieties in each dimension. There is another well-known conjecture asserting that the anticanonical degree of a Fano variety is bounded from above by a universal constant depending only on the dimension. These two boundedness conjectures are equivalent, by [KoMa] or [De].

Boundedness of Fano varieties of dimension three or less follows from classification. Indeed, the only Fano 1-fold is $\mathbf{P}^1$; the only Fano 2-folds are the del Pezzo surfaces; and the classification of Fano 3-folds by Fano, Iskovskikh, Mori, Mukai, and Shokurov implies that there are precisely 104 deformation types of Fano 3-folds. Boundedness of toric Fano varieties was established in [Ba]. Boundedness of Fano varieties of the first kind was established in [Na1] in dimension four, and in [KoMiMo] and [Na2] in arbitrary dimensions. For a differential-geometric approach to boundedness in the case of rationally-connected Fano varieties, see [Ts], and for a generalization of [Na1, Na2] from the Fano case to the case of arbitrarily-polarized varieties covered by free-moving curves of arbitrary genus, see [Ca].

Here we shall be interested in the following setup. Let $\pi : M \to X$ be a connected (i.e., having connected fibers) surjective morphism from a (smooth) Fano variety M onto a (possibly singular) projective variety X. Let L be an ample (actually, nef and big is enough) line bundle on X. Let F be a general fiber of our morphism; it too is a Fano variety. Let k be the least integer such that K_F^{-k} is very ample on F; according to [De], $k \leq 12f^f$. Set $m = \dim M$, $f = \dim F$, and $x = \dim X$. Our main result is as follows.

Theorem A. *In the above situation, we have the estimate* $c_1(M)^m \leq 3(m\tau)^m \mid (X, L) \mid$, *where* $\mid (X,L) \mid = \min_{0 \leq \eta \leq x} h^0(X, L^\eta)$, *and where where* $\tau = k^{f-1}c_1(F)^f + 2$ *if* $f > 0$ *and* $\tau = 2$ *if* $f = 0$.

Of course, what we would ultimately like is a "universal estimate" – that is, an estimate of $c_1(M)^m$ in terms of m alone. What Theorem A provides is a "relative estimate" – that is, an estimate of $c_1(M)^m$ in terms of the "algebro-geometric complexity" of the base X and general fiber F. This suggests an inductive approach to proving boundedness, using induction on dimension or Picard number. One of the main difficulties with this approach is that the base X may conceivably be very singular. In some situations, we may already have at our disposal universal estimates for the base and general fiber, and may then be able to use Theorem A to obtain universal estimates for $c_1(M)^m$. As an illustration, suppose M is any Fano 4-fold that admits a nontrivial morphism $M \to C$ to a curve. Assume for simplicity that K_M^{-1} is already very ample. We shall obtain a universal estimate for $c_1(M)^4$. Since Fano varieties always have irregularity zero, the curve must be $\mathbf{P}^1$. Using Stein factorization, we may assume the morphism to be connected. We thus have a connected surjective morphism $M \to \mathbf{P}^1$. The general fiber F is a Fano 3-fold, and by classification of Fano 3-folds we have $c_1(F)^3 \leq 64$. We can take $k = 1$ because K_M^{-1} is very ample by assumption and $K_F^{-1} = K_M^{-1} \mid_F$. We thus have $\tau \leq 66$. Clearly $\mid (\mathbf{P}^1, \mathcal{O}_{\mathbf{P}^1}(1)) \mid = 2$, so by Theorem A we obtain the universal bound $c_1(M)^4 \leq 3 \cdot (4 \cdot 66)^4 \cdot 2$, as desired.

This article is organized as follows. The first section contains material on coherent sheaves of ideals associated to almost-plurisubharmonic functions, and culminates in a cohomology vanishing theorem. The second section contains various bits and pieces, mainly in the form of lemmas. And the third section contains the actual proof of Theorem A.

I would like to take this opportunity to thank the Taniguchi foundation and the organizers of the Taniguchi conference for enabling me to participate and for making my experience very rewarding. I would also like to thank the other participants.

1 Coherent sheaves of ideals associated to almost-plurisubharmonic functions

Following [Na1], we present a simplified version of the theory of multiplier ideal sheaves [Na3, Na4] that deals with individual singular almost-plurisubharmonic functions rather than sequences of smooth functions. See also [De].

1.1 The almost-plurisubharmonic function φ

Definition 1.1 *A function* $\varphi : M \to \mathbf{R} \cup \{\infty\}$ *on a complex manifold* M *is said to be* **almost-plurisubharmonic** *if it is locally expressible as the sum of a plurisubharmonic function and a smooth function.*

1.2 The coherent sheaf of ideals $\mathcal{I}(\varphi)$

Consider an almost-plurisubharmonic function φ on a complex manifold M. We define a coherent analytic sheaf of ideals $\mathcal{I}(\varphi)$ on M as follows. For each nonempty open subset $U \subset M$ we let $\Gamma(U, \mathcal{I}(\varphi))$ be the set of all holomorphic functions f on U such that $| f |^2 e^{-\varphi}$ is locally summable on U. Clearly $\Gamma(U, \mathcal{I}(\varphi))$ is an ideal in the ring $\Gamma(U, \mathcal{O}_M)$. Because local summability is a local notion, we obtain an analytic sheaf of ideals and not merely a presheaf. Coherence will be proved shortly, but first we record the following obvious fact.

Proposition 1.1 *For every smooth function* $\psi : M \to \mathbf{R}$ *we have* $\mathcal{I}(\varphi) = \mathcal{I}(\varphi + \psi)$.

Bombieri and Hörmander have shown that the exponential of any plurisuperharmonic function is locally summable at some point. Consequently, we have the following.

Proposition 1.2 $\mathcal{I}(\varphi)$ *is not identically zero.*

We turn next to the question of coherence.

Theorem 1.1 (Coherence) *For any almost-plurisubharmonic function φ on any complex manifold M, the analytic sheaf of ideals $\mathcal{I}(\varphi)$, as defined above, will be coherent.*

The proof will require several preparatory lemmas.

Lemma 1.1 *Suppose φ is an almost-plurisubharmonic function on a Stein manifold M. For any nonempty open subset $U \subset M$, any holomorphic function $f \in \Gamma(U, \mathcal{I}(\varphi))$, any point $p \in U$, and any positive integer k, there exists a global holomorphic function $g \in \Gamma(M, \mathcal{I}(\varphi))$ such that $f - g$ vanishes to order greater than k at the point p.*

This lemma is proved by using L^2 estimates of $\overline{\partial}$ with respect to the weight function $\varphi + \psi$ where ψ is a suitable plurisubharmonic function on M that is smooth except at p, where it has a substantial pole. The following lemma can be found in most standard texts on algebra.

Lemma 1.2 (Krull) *Suppose R is a Noetherian local ring with maximal ideal J. Then for any ideal $I \subset R$ we have*

$$I = \bigcap_{k=1}^{\infty} (I + J^k).$$

Lemma 1.3 *Let F be an arbitrary collection of global holomorphic functions on a complex manifold M. Then the sheaf of ideals generated by F is coherent.*

In the event that F has finite cardinality, this is a consequence of Oka's coherence theorem. In the general case, one uses the fact that any ascending chain of coherent sheaves of ideals eventually stabilizes over each compact subset, to reduce to the special case.

Proof of theorem. As coherence is a local notion, we can assume M to be Stein. Let $\mathcal{F}$ denote the analytic sheaf of ideals on M generated by the global holomorphic functions in $\Gamma(M, \mathcal{I}(\varphi))$. By Lemma 1.3, we know $\mathcal{F}$ is coherent. Clearly $\mathcal{F} \subset \mathcal{I}(\varphi)$. To complete the proof, we will demonstrate the reverse inclusion. Fix any point $p \in M$. By Lemma 1.1 we have

$$\mathcal{I}(\varphi)_p \subset \bigcap_{k=1}^{\infty} (\mathcal{F}_p + m_p^k)$$

where m_p denotes the maximal ideal of the local ring $\mathcal{O}_{M,p}$. By Lemma 1.2 we conclude that $\mathcal{I}(\varphi)_p \subset \mathcal{F}_p$, and the theorem is proved. **QED**

1.3 The complex analytic subspace $V(\varphi) \subset M$

Again consider an almost-plurisubharmonic function φ on a complex manifold M. Denote by $V(\varphi) \subset M$ the (possibly nonreduced or empty) complex analytic subspace cut out by the coherent sheaf of ideals $\mathcal{I}(\varphi) \subset \mathcal{O}_M$. Observe that, as a subset, $V(\varphi)$ consists of precisely those points at which $e^{-\varphi}$ fails to be locally summable.

1.4 A fine resolution for $\mathcal{I}(\varphi)$

Consider once again an almost-plurisubharmonic function φ on a complex manifold M. Set $m = \dim M$. In the preceding paragraphs we constructed a coherent sheaf of ideals $\mathcal{I}(\varphi)$. In the following paragraphs we shall construct a fine resolution of $\mathcal{I}(\varphi)$, a twisted variant of which will subsequently be used to prove a cohomology vanishing theorem for $\mathcal{I}(\varphi)$.

Define for each $q \in \{0, 1, \ldots, m\}$ a sheaf $\mathcal{L}^{0,q}(\varphi)$ as follows. For each nonempty open subset $U \subset M$ let $\Gamma(U, \mathcal{L}^{0,q}(\varphi))$ be the set of all $(0, q)$-forms f on U for which the following two conditions are satisfied.

1. The coefficients of f are measurable, and $|f|^2 e^{-\varphi}$ is locally summable (on U).

2. There exists a $(0, q+1)$-form g on U with measurable coefficients such that $\bar{\partial} f = g$ in the sense of distributions (on U) and such that $|g|^2 e^{-\varphi}$ is locally summable (on U).

(Note: The norm $|f|$ is taken with respect to any smooth Riemannian metric on M. The particular choice of metric is clearly inessential.) As these conditions are essentially local in nature, we obtain a sheaf and not merely a presheaf.

We note that, for any $f \in \Gamma(U, \mathcal{L}^{0,q}(\varphi))$ and any smooth function ρ on U, the product ρf also belongs to $\Gamma(U, \mathcal{L}^{0,q}(\varphi))$. The essential point here is that the Leibniz product formula $\bar{\partial}(\rho f) = (\bar{\partial}\rho) f + \rho(\bar{\partial} f)$ holds. Consequently, the sheaf $\mathcal{L}^{0,q}(\varphi)$ is **fine**.

Lemma 1.4 (Poincare lemma) *For any Stein open subset $U \subset M$, any $q \in \{0, 1, \ldots, m\}$, and any $g \in \Gamma(U, \mathcal{L}^{0,q+1}(\varphi))$ with $\bar{\partial} g = 0$, there exists $f \in \Gamma(U, \mathcal{L}^{0,q}(\varphi))$ such that $\bar{\partial} f = g$.*

This is a consequence of Hörmander's L^2 estimates of $\bar{\partial}$, in the case of singular weight functions. The above discussion can be summarized as follows.

Theorem 1.2 (Fine resolution) *For any almost-plurisubharmonic function φ on any complex manifold M,*

$$0 \to \mathcal{I}(\varphi) \to \mathcal{L}^{0,0}(\varphi) \xrightarrow{\overline{\partial}} \cdots \xrightarrow{\overline{\partial}} \mathcal{L}^{0,m}(\varphi) \to 0$$

is an exact sequence of sheaves on M, and provides a fine resolution of the coherent analytic sheaf of ideals $\mathcal{I}(\varphi)$. Consequently, the cohomology groups $H^*(M, \mathcal{I}(\varphi))$ can be identified with the cohomology groups of the following complex of global sections:

$$0 \to \Gamma(M, \mathcal{L}^{0,0}(\varphi)) \to \cdots \to \Gamma(M, \mathcal{L}^{0,m}(\varphi)) \to 0.$$

1.5 A cohomology vanishing theorem for $\mathcal{I}(\varphi)$

Suppose φ is an almost-plurisubharmonic function on a projective algebraic Kähler manifold M, and L is a Hermitian holomorphic line bundle on M. Denote by ω the Kähler form on M, and by $c(L)$ the curvature form of L. Our cohomology vanishing theorem is as follows.

Theorem 1.3 (Cohomology Vanishing Theorem) *Assume the above setup, and suppose there exists $\epsilon > 0$ such that*

$$\partial\overline{\partial}\,\varphi + Ric(\omega) + c(L) \geq \epsilon\omega$$

on M. Then $H^i(M, \mathcal{I}(\varphi) \otimes L) = 0$ for $i > 0$.

For the proof we use the fine resolution of $\mathcal{I}(\varphi)$ introduced above, but we twist it by L. Fix $q \in \{0, 1, \ldots, m\}$ and $g \in \Gamma(M, \mathcal{L}^{0,q+1}(\varphi) \otimes L)$ with $\overline{\partial} g = 0$. We can use Hörmander's L^2 estimates for $\overline{\partial}$ in conjunction with the Bochner-Kodaira formula to obtain $f \in \Gamma(M, \mathcal{L}^{0,q}(\varphi) \otimes L)$ such that $\overline{\partial} f = g$. The theorem follows.

2 Other preparatory material

2.1

In this subsection we shall apply material from the previous section to the Fano situation in a form that will be needed in the proof of Theorem A. Let M be a Fano variety, $\nu > 0$

a positive integer, $s \in H^0(M, K_M^{-\nu}) - \{0\}$ a nontrivial pluri-anticanonical section, and L a semipositive Hermitian holomorphic line bundle on M. Fix a Kähler form ω on M, and fix a Hermitian metric along the fibers of K_M^{-1} (to be specified more precisely later). Choose any real number $\nu' > \nu$ and set $\varphi = \frac{1}{\nu'} \log \mid s \mid^2$. Our goal will be to show that $H^i(M, \mathcal{I}(\varphi) \otimes L) = 0$ for $i > 0$.

We have $\partial\overline{\partial}\ \varphi + \frac{\nu}{\nu'} c(K_M^{-1}) \geq 0$ essentially by the the definition of φ, and $c(L) \geq 0$ by hypothesis. Hence $\partial\overline{\partial}\ \varphi + \mathrm{Ric}(\omega) + c(L) \geq \mathrm{Ric}(\omega) - \frac{\nu}{\nu'} c(K_M^{-1})$. Now the form $\mathrm{Ric}(\omega) - \frac{\nu}{\nu'} c(K_M^{-1})$ represents the cohomology class $(1 - \frac{\nu}{\nu'})c_1(M)$, which is positive, and hence will itself be pointwise positive provided we choose the Hermitian metric on K_M^{-1} appropriately. Hence the cohomology vanishing theorem from the previous section applies, and we get $H^i(M, \mathcal{I}(\varphi) \otimes L) = 0$ for $i > 0$, as desired.

2.2 Local integrability

Let f be a holomorphic function defined locally near the origin 0 in $\mathbf{C}^m$. Let τ be a positive real number and set $\varphi = \frac{1}{\tau} \log \mid f \mid^2$. If $\tau > \mathrm{mult}_0(f)$ then $e^{-\varphi}$ will be locally integrable at 0, while if $\tau < \mathrm{mult}_0(f)/m$ then $e^{-\varphi}$ will not be locally integrable at 0.

2.3 Three lemmas

Lemma 2.1 *Let M be a complex manifold, L a holomorphic line bundle on M, and $r > 0$ an integer. Assume that, for any choice of r points $p_1, \ldots, p_r \in M$, there exists a section $s \in H^0(M, L) - \{0\}$ vanishing at all the p_i. Then $h^0(M, L) > r$.*

Proof of lemma. Each time we "assign" a new base point to $H^0(M, L)$, its dimension drops by exactly one. This is valid as long as the new assigned base point is not already an existing base point. Thanks to the hypothesis, we can continue assigning new base points, one at a time, for a total of $r + 1$ times. The result follows. **QED**

Lemma 2.2 *Let M be an m-dimensional Fano variety, and let k be an integer such that K_M^{-k} is very ample. For any integer $\nu > 0$, any section $s \in H^0(M, K_M^{-\nu}) - \{0\}$, and any point $p \in M$, we have the estimate $\mathrm{mult}_p(s) \leq \nu k^{m-1} c_1(M)^m$.*

Proof of lemma. Use the line bundle K_M^{-k} to embed M into a projective space, as a subvariety of degree $k^m c_1(M)^m$. Take complete intersections with linear subspaces to get a curve $C \subset M$ passing through p in a general direction. The projective degree of C will be $k^m c_1(M)^m$; the anticanonical degree of C (i.e., $-K_M \cdot C$) will be $k^{m-1} c_1(M)^m$. Now restrict s to C to obtain the desired estimate. **QED**

Lemma 2.3 (Siegel-type lemma) *Let M be an m-dimensional connected compact complex manifold, let L be a positive holomorphic line bundle on M, and let r be a positive integer. For any choice of r distinct points $p_1, \ldots, p_r \in M$ and any real number $A < (\frac{1}{r} c_1(L)^m)^{\frac{1}{m}}$, there exist an integer $\nu > 0$ and a section $s \in H^0(M, L^\nu) - \{0\}$ whose vanishing orders at the p_i exceed νA.*

Proof of lemma. By Riemann-Roch and Serre vanishing we have

$$h^0(M, L^\nu) \geq \frac{1}{m!} c_1(L)^m \nu^m + O(\nu^{m-1})$$

as $\nu \to \infty$. On the other hand,

$$h^0(M, \mathcal{O}_M / \mathcal{I}^{\eta+1}) = r\binom{m+\eta}{m} \leq \frac{r}{m!} \eta^m + O(\eta^{m-1})$$

as $\eta \to \infty$. Here $\mathcal{I}$ denotes the ideal sheaf of the set $p_1, \ldots, p_r$. Now set $\eta = [\nu A]$ for some $\nu >> 0$. Here the brackets denote the greatest integer function. The above inequalities, together with the inequality $A < (\frac{1}{r} c_1(M)^m)^{\frac{1}{m}}$, imply that $h^0(M, L^\nu) > h^0(M, \mathcal{O}_M / \mathcal{I}^{\eta+1})$. Hence there exists a section $s \in H^0(M, L^\nu) - \{0\}$ whose Taylor series expansions of order η at the p_i is trivial. **QED**

3 Proof of Theorem A

In this section we shall prove Theorem A. We have not attempted to make the various estimates sharp, and there is considerable room for improvement. Let $\pi : M \to X$, F, L, k, m, f, x be as in the setup of Theorem A. Let $| (X, L) | = \min_{0 \leq \eta \leq x} h^0(X, L^\eta)$, and let $\tau = k^{f-1}c_1(F)^f + 2$ if $f > 0$ and $\tau = 2$ if $f = 0$. Our goal is to show that $c_1(M)^m \leq 3(m\tau)^m \mid (X, L) \mid$.

3.1

If $c_1(M)^m \leq 3(m\tau)^m$ then we are done. Now assume that $c_1(M)^m > 3(m\tau)^m$, and set

$$r = [c_1(M)^m / 2(m\tau)^m] \geq 1.$$

Here the brackets signify greatest integer. Pick any r points $p_1, \ldots, p_r \in M$.

3.2

According to the Siegel-type lemma, there exist an integer $\nu > 0$ and a section $s \in H^0(M, K_M^{-\nu}) - \{0\}$ such that

$$\text{mult}_{p_i}(s) > \nu(\frac{1}{2r}c_1(M)^m)^{\frac{1}{m}}, \quad (1 \leq i \leq r).$$

3.3

(If $f = 0$ then this paragraph becomes trivial and should be skipped.) We may assume that the fiber F is in general position relative to the section s, so that $s \mid_F$ is not identically zero. Then by Lemma 2.2, the vanishing multiplicity of $s \mid_F$ at each point in F is bounded from above by $\nu(\tau - 1)$.

3.4

Set $\varphi = \frac{1}{\nu\tau} \log \mid s \mid^2$. (The norm $\mid s \mid$ is taken with respect to any Hermitian metric along the fibers of K_M^{-1}.) By Subsection 2.1 we get the cohomology vanishing result $H^i(M, \mathcal{I}(\varphi) \otimes \pi^* L^\eta) = 0$ for $i \geq 1$ and $\eta \geq 0$.

3.5

From the lower estimate (3.2) on the multiplicity of s at the p_i, we conclude that $e^{-\varphi}$ is not locally summable at the p_i. (See Subsection 2.2.) That is, $\{p_1, \ldots, p_r\} \subset V(\varphi)$.

3.6

From the upper estimate (3.3) on the multiplicity of s along F, we conclude that $e^{-\varphi}$ is locally summable along F. (See Subsection 2.2.) That is, $V(\varphi) \cap F = \emptyset$.

3.7

The image $\pi(V(\varphi)) \subset X$ is not equal to all of X, by (3.6). Since L is ample on X, some power L^η admits a nontrivial section that vanishes on $\pi(V(\varphi))$. This implies that $h^0(M, \mathcal{I}(\varphi) \otimes \pi^* L^\eta) \neq 0$ for some η. On the other hand, cohomology vanishing (3.4) gives $h^i(M, \mathcal{I}(\varphi) \otimes \pi^* L^\eta) = 0$ for all $i \geq 1$ and $\eta \geq 0$ (see (3.4)). Now the Euler characteristic $\chi(M, \mathcal{I}(\varphi) \otimes \pi^* L^\eta)$ is a polynomial of degree at most x in η, and is nontrivial by the above considerations. Hence it must be nonzero for some $\eta \in \{0, \ldots, x\}$. Therefore $h^0(M, \mathcal{I}(\varphi) \otimes \pi^* L^\eta) \neq 0$ for some $\eta \in \{1, \ldots, x\}$.

3.8

From (3.5) and (3.7) we know that $H^0(M, \pi^* L^\eta)$ contains a nontrivial element that vanishes at all the p_i; this is true for some $\eta \in \{1, \ldots, x\}$. Since the p_i were arbitrarily chosen, we can apply Lemma 2.1 to get the estimate $h^0(M, \pi^* L^\eta) > r$. (There is one point here that we need to be careful about. Namely, different choices of the p_i may lead to different $\eta \in \{0, \ldots, x\}$. However, this doesn't cause any concern, since $\{1, \ldots, x\}$ is finite.) Equivalently, $h^0(X, L^\eta) > r$. Finally, we need only recall the definitions of $| (X, L) |$ and r to see that this gives the inequality $c_1(M)^m \leq 2(m\tau)^m | (X, L) |$, as desired. This completes the proof of Theorem A.

References

[Ba] V. V. Batyrev, Boundedness of the degree of multidimensional toric Fano varieties, Moscow Univ. Math. Bull. **37** (1982), 28-33.

[Ca] F. Campana, Une version géométrique généralisée du théorème du produit de Nadel, preprint.

[De] J.-P. Demailly, A numerical criterion for very ample line bundles, preprint.

[KoMa] J. Kollár and T. Matsusaka, Riemann-Roch type inequalities, Amer. J. Math. **105** (1983), 229-252.

[KoMiMo] J. Kollár, Y. Miyaoka, and S. Mori, Rational curves in Fano varieties, preprint.

[Na1] A. M. Nadel, A finiteness theorem for Fano 4-folds, preprint.

[Na2] A. M. Nadel, The boundedness of degree of Fano varieties with Picard number one, Journal of the A.M.S. (in press).

[Na3] A. M. Nadel, Multiplier ideal sheaves and existence of Kähler-Einstein metrics of positive scalar curvature, *Proc. Nat. Acad. Sci. U.S.A.* **86** (1989), 7299-7300.

[Na4] A. M. Nadel, Multiplier ideal sheaves and Kähler-Einstein metrics of positive scalar curvature, Annals of Math. **132** (1990), 549-596.

[Ts] H. Tsuji, Boundedness of the degree of Fano manifolds with $b_2 = 1$, preprint.

14

Monopoles and Nahm's Equations

Hiraku Nakajima*

Department of Mathematics, University of Tokyo
Hongo 7-3-1, Tokyo 113, Japan**

Abstract. We give a different proof of Hitchin's result : a correspondence between SU(2)-monopoles and solutions of Nahm's equations. We also prove that this correspondence gives a hyper-Kähler isometry between the monopole moduli space and the space of equivalence classes of solutions of Nahm's equations, equipped with their natural metrics. Such a result was conjectured by Atiyah and Hitchin.

1. Introduction

In 1983 Hitchin [Hi3] gave an equivalence between

A) an SU(2) monopole satisfying certain asymptotic conditions,

B) a solution of Nahm's equation satisfying certain boundary conditions.

The correspondence B $\Rightarrow$ A is an adaptation of the Atiyah-Drinfeld-Hitchin-Manin construction [ADHM] of instantons on S^4, and was produced by Nahm [Na1]. Hitchin constructed the correspondence A $\Rightarrow$ B by relating A and B to the third object:

C) a compact algebraic curve in $T\mathbb{P}^1$ satisfying certain conditions.

The third object C is interesting to explore in itself, but for the purpose in giving the correspondence A $\Rightarrow$ B, this approach is indirect and it is not so easy to prove that the composition A $\Rightarrow$ B $\Rightarrow$ A gives back the same monopole.

Later Nahm [Na2] and Corrigan-Goddard [CG] pointed out a new approach which is more direct. From their point of view, the transform which produces B $\Rightarrow$ A and A $\Rightarrow$ B can be considered as analogous to a Fourier transform, so it seems very natural, at least philosophically, that two correspondences are mutually inverse. But they do not check the

*Partially supported by Grant-in-Aid for Scientific Reserch (No. 02854001), Ministry of Education, Science and Culture, Japan.

**Current Address: Mathematical Insititute, Tôhoku University, Aramaki, Aoba-ku, Sendai 980, Japan

boundary behaviour of the solutions of Nahm's equations. This is the remaining part in their approach.

Our aim is to fill the hole in their approach. But for the sake of the reader, we shall give the proofs (sometimes only in outline) in the whole steps.

Let us give the precise statement of Hitchin's result. Our objects are the following:

A) An SU(2) connection A on a rank 2 hermitian vector bundle E over $\mathbb{R}^3$ and a skew hermitian endomorphism Φ (the Higgs field) satisfying

A1) (the Bogomolny equation)

$$*R_A = d_A\Phi,$$

A2) the asymptotic expansion as $r = |x| \to \infty$, up to gauge transformation,

$$\Phi = \begin{pmatrix} i(1-\frac{k}{2r}) & 0 \\ 0 & -i(1-\frac{k}{2r}) \end{pmatrix} + O(r^{-2}),$$

$$|\nabla_A\Phi| = O(r^{-2}), \qquad \frac{\partial|\Phi|}{\partial\Omega} = O(r^{-2}),$$

where k is a positive integer.

B) A hermitian connection ∇ on a hermitian vector bundle V of rank k over the open interval $I = (-1,1)$ and three skew-hermitian endomorphisms $T_\alpha \in \Gamma(I; \mathrm{Endskew}(V))$ satisfying

B1) (the Nahm's equation)

$$\nabla_t T_\alpha + \frac{1}{2}\sum_{\beta,\gamma}\varepsilon_{\alpha\beta\gamma}[T_\beta, T_\gamma] = 0,$$

B2) T_α has at most simple poles at $t = \pm 1$ but is otherwise analytic,

B3) at each pole the residues of (T_1, T_2, T_3) define an irreducible representation of $\mathfrak{su}(2)$. Namely near the endpoint $t = 1$, in a covariant constant basis, we can write

$$T_\alpha(t) = \frac{a_\alpha}{t-1} + b_\alpha(t),$$

where b_α is analytic in a neighbourhood of $t = 1$. Then

$$x_1e_1 + x_2e_2 + x_3e_3 \mapsto -2(x_1a_1 + x_2a_2 + x_3a_3)$$

defines a k-dimensional representation of $\mathfrak{su}(2)$. (This is a consequence of the Nahm's equation.) Here (e_1, e_2, e_3) is a basis for $\mathfrak{su}(2)$ defined by

$$e_1 = \begin{pmatrix} i & 0 \\ 0 & -i \end{pmatrix}, \quad e_2 = \begin{pmatrix} 0 & -1 \\ 1 & 0 \end{pmatrix}, \quad e_3 = \begin{pmatrix} 0 & i \\ i & 0 \end{pmatrix}.$$

The last condition says this representation is *irreducible*, and similarly at the other pole $t = -1$.

Now our main result is

Theorem (Hitchin [Hi3]). *There is a natural equivalence between monopoles satisfying conditions* A *and Nahm data satisfying conditions* B.

The formal aspects of the proof is the same as the instanton case (see [Na2, CG]). In this case the similar proof but incorporating the complex geometry was given by Donaldson in [DK]. These two methods were used and presented side-by-side in [KN]. We shall adapt Nahm-Corrigan-Goddard's method in principle, but use the complex notation hoping that it makes the calculation familiar.

The paper is organized as follows. In Sect. 2 we give the correspondence A $\Rightarrow$ B. In Sect. 3 we review the construction of monopoles from Nahm data, i.e., B $\Rightarrow$ A. In Sects. 4 and 5 we prove that two correspondences are mutually inverse. Section 6 takes up an interesting side-issue: we show that our correspondence gives a hyper-Kähler isometry between the space of equivalence classes of solutions of Nahm's equations and the moduli space of monopoles, equipped with their natural hyper-Kähler structures. This was conjectured by Atiyah and Hitchin [AH]. Section 7 provides some remarks. In the appendix we shall give a proof of Lemma which we need in Sect. 2.

In a future work, the author hopes to extend Main Theorem to SU(m)-monopoles. The only thing left is to study the boundary behaviour of the solutions of Nahm's equations. There are results of [HM] in this direction.

2. From Monopoles to Nahm's equations

The purpose of this section is to obtain from an SU(2)-monopole (A, Φ) a solution T_α to Nahm's equations which satisfies the conditions B of the introduction.

For each $t \in I$ consider the following operators:

$$D_{A,t} = D_A + (\Phi - it) \colon \Gamma(S \otimes E) \to \Gamma(S \otimes E),$$
$$D^*_{A,t} = D_A - (\Phi - it) \colon \Gamma(S \otimes E) \to \Gamma(S \otimes E),$$

where S is the spin bundle over $\mathbb{R}^3$ and D_A is the Dirac operator coupled with the connection A. Note that $D^*_{A,t}$ is the formal adjoint of $D_{A,t}$. Then the Weitzenböck formula shows

$$D^*_{A,t} D_{A,t} = 1_S \otimes (\nabla^*_A \nabla_A - (\Phi - it)^2), \tag{2.1}$$

which is a positive operator. In particular, $D_{A,t}$ has no L^2 kernel. Define

$$V_t = L^2 \text{ kernel of } D^*_{A,t}.$$

By an index theorem [Ca] the index of the operator $D_{A,t}$ is equal to $-k$, and we have $\operatorname{Ker} D_{A,t} = 0$. So V_t defines a vector bundle V of rank k on I which is a sub-bundle of the trivial bundle $\underline{L^2(\mathbb{R}^3; S \otimes E)}$ over I. (We denote by $\underline{W}$ the trivial bundle whose fiber is a vector space W.) Let π be the orthogonal projection onto V. Define a connection and three endomorphisms on V by

$$\nabla_t \psi = \pi\left(\frac{\partial \psi}{\partial t}\right), \qquad T_\alpha(\psi) = \pi(i x_\alpha \psi), \qquad \alpha = 1, 2, 3.$$

Note that $i x_\alpha \psi$ is in L^2 since ψ decays exponentially as $r = |x| \to \infty$.

The skew-hermiticity of T_α is automatic from the definition of T_α. So we first study the boundary condition B2. Since the behaviour when $t \to +1$ is similar, we only study the case $t \to -1$.

In the following calculation, we use the constant C in the generic sense. So the symbol C may mean different constants in different equations. The important point is that C must be independent of t, since we want to study the behaviour as $t \to -1$.

As shown in [Hi2,p.591], under the condition A1, there is an asymptotic gauge in which the Higgs field is the form of B2 and the connection matrix has the following asymptotic behaviour:

$$\begin{pmatrix} A_0^* & 0 \\ 0 & A_0 \end{pmatrix} + O(r^{-2}),$$

where A_0 is the connection form for a homogeneous connection on a line bundle of degree k over $S^2 = \mathbb{P}^1$ extended radially to $\mathbb{R}^3 \setminus \{0\}$ (A_0^* is its dual). Take a radial coordinate system (r, θ). The spinor bundle of $S^2 \times (0, \infty)$ is isomorphic to that of S^2, hence decomposes as $S^+ \oplus S^-$. Then there exists a norm preserving bundle isomorphism between S and $S^+ \oplus S^-$ under which Dirac operators are related by (cf. [Hi1])

$$D\psi = r^{-2} D_{S^+ \oplus S^-}(r\psi) = \begin{pmatrix} i(\frac{\partial}{\partial r} + \frac{1}{r}) & \frac{1}{r} D^- \\ \frac{1}{r} D^+ & -i(\frac{\partial}{\partial r} + \frac{1}{r}) \end{pmatrix},$$

where $D^\pm$ is the Dirac operator on S^2. Let denote the Dirac operators on S^2 twisted by A_0, A_0^* by $D^\pm_{A_0}$, $D^\pm_{A_0^*}$. The operator $D^*_{A,t}$ can be represented as follows:

$$D^*_{A,t} = \begin{pmatrix} B_1 & 0 \\ 0 & B_2 \end{pmatrix} + O(r^{-2}),$$

where

$$B_1 = \begin{pmatrix} i(\frac{\partial}{\partial r} + t - 1 + \frac{k+2}{2r}) & \frac{1}{r} D^-_{A_0^*} \\ \frac{1}{r} D^+_{A_0^*} & -i(\frac{\partial}{\partial r} - t + 1 - \frac{k-2}{2r}) \end{pmatrix},$$

$$B_2 = \begin{pmatrix} i(\frac{\partial}{\partial r} + t + 1 - \frac{k-2}{2r}) & \frac{1}{r} D^-_{A_0} \\ \frac{1}{r} D^+_{A_0} & -i(\frac{\partial}{\partial r} - t - 1 + \frac{k+2}{2r}) \end{pmatrix}.$$

Let R be a fixed positive number and χ a cut-off function which is 0 on $[0, R]$ and 1 on $[R+1, \infty)$. Using the isomorphisms $S^+ \cong \Lambda^{0,0} \otimes H^*$, $S^- \cong \Lambda^{0,1} \otimes H^*$ (where H is a hyperplane bundle), we can define $\psi \in L^2(\mathbb{R}^3; S \otimes E)$ from $f \in H^0(\mathbb{P}^1; \mathcal{O}(k-1))$ by

$$\psi(r, \theta) = \begin{pmatrix} 0 & 0 & \chi(r) e^{-(t+1)r} r^{\frac{k-2}{2}} f(\theta) & 0 \end{pmatrix}^t.$$

Then it satisfies

$$|D^*_{A,t}\psi|_S \le C e^{-(t+1)r} (r+1)^{\frac{k-2}{2} - 2},$$

for some constant C depending only on (A, Φ), χ and $\sup|f|$. This means that ψ is an approximate solution of $D^*_{A,t}\psi = 0$. A real solution is given by $\psi - D_{A,t}\varphi$ where φ is the unique solution

$$D^*_{A,t} D_{A,t} \varphi = 1_S \otimes (\nabla_A^* \nabla_A - (\Phi - it)^2)\varphi = D^*_{A,t}\psi. \tag{2.2}$$

We shall show that $D_{A,t}\varphi$ is small relative to ψ, so the boundary behaviour of T_α is determined from ψ.

The equation (2.2) can be uniquely solved by the same method as in [JT, Proposition IV.4.1]. Please see [JT] for details. The solution φ is the minimum of the functional

$$S(\varphi) = \|\nabla_A\varphi\|_{L^2}^2 + \|(\Phi - it)\varphi\|_{L^2}^2 - 2\langle \varphi, D_{A,t}^*\psi\rangle_{L^2}.$$

This is strictly convex, differential and coercive, so has a unique minimum. In particular, $S(\varphi) \le S(0) = 0$. Hence,

$$\|D_{A,t}\varphi\|_{L^2}^2 = \|\nabla_A\varphi\|_{L^2}^2 + \|(\Phi - it)\varphi\|_{L^2}^2 \le 2\langle \varphi, D_{A,t}^*\psi\rangle_{L^2}. \tag{2.3}$$

If R is sufficiently large and t is near -1, we have an estimate

$$(1+t)|\varphi| \le 2|(\Phi - it)\varphi| \qquad \text{in } \mathbb{R}^3 \setminus B_{\frac{R}{1+t}}.$$

So we get

$$(1+t)^2 \int_{\mathbb{R}^3\setminus B_{\frac{R}{1+t}}} |\varphi|^2\, dx \le 4\|(\Phi - it)\varphi\|_{L^2}^2. \tag{2.4}$$

On the other hand, the integral over $B_{\frac{R}{1+t}}$ can be estimated by using the Hölder's and Sobolev inequalities as

$$(1+t)^2 \int_{B_{\frac{R}{1+t}}} |\varphi|^2\, dx \le C\|\varphi\|_{L^6}^2 \le C\|\, d|\varphi|\, \|_{L^2}^2 \le C\|\nabla_A\varphi\|_{L^2}^2, \tag{2.5}$$

where we have used the Kato's inequality in the last step. Substituting (2.4) and (2.5) into (2.3), we get

$$\|D_{A,t}\varphi\|_{L^2} \le C(1+t)^{-1}\|D_{A,t}^*\psi\|_{L^2}$$

Direct calculation shows that

$$\|D_{A,t}^*\psi\|_{L^2} \le C(1+t)^2\|\psi\|_{L^2}.$$

Hence $D_{A,t}\varphi$ is small if t is sufficiently near to -1, as required.

Thus we have obtained a trivialization of the bundle V near $t = -1$ (after the Gram-Schmidt orthogonalization). This trivialization is not covariant constant, but the trace-free part of the connection form is bounded. Hence it is enough to study the asymptotic behaviour in this trivialization. So the condition B3 follows from

Lemma 2.6. *Let a_α be an endomorphism of $H^0(\mathbb{P}^1; \mathcal{O}(k-1))$ defined by*

$$\langle a_\alpha f_1, f_2\rangle = \int_{\mathbb{P}^1} \langle i x_\alpha f_1, f_2\rangle\, dV.$$

Then a non-zero constant multiple of a linear map $x_1e_1 + x_2e_2 + x_3e_3 \mapsto x_1a_1 + x_2a_2 + x_3a_3$ defines an irreducible k-dimensional representation of $\mathfrak{su}(2)$.

The proof will be given in the Appendix. Remark that we will prove that T_α's satisfy the Nahm's equations below, so the constant must be equal to -2.

Proposition 2.7. *The endomorphisms T_α and the connection ∇ satisfy the Nahm's equations*

$$\nabla_t T_\alpha + \frac{1}{2}\sum_{\beta,\gamma} \varepsilon_{\alpha\beta\gamma}[T_\beta, T_\gamma] = 0, \qquad \alpha = 1,2,3.$$

Before entering the proof of this proposition, we prepare the complex notation as in [Do] by breaking the natural symmetry and choosing a particular isomorphism $\mathbb{R}^3 \cong \mathbb{R} \times \mathbb{C}$.

Fixing a trivialization of the bundle V, we write the connection ∇ as $\frac{d}{dt} + T_0$. Put

$$\alpha = \frac{1}{2}(T_0 + iT_1), \qquad \beta = \frac{1}{2}(T_2 + iT_3).$$

Then the Nahm's equations become the following pair of equations:

$$\frac{d\beta}{dt} + 2[\alpha, \beta] = 0 \qquad \text{(the complex equation)},$$
$$\frac{d}{dt}(\alpha + \alpha^*) + 2([\alpha, \alpha^*] + [\beta, \beta^*]) = 0 \qquad \text{(the real equation)}.$$

To prove that T_α's satisfy the Nahm equations, it is not necessarily to check both the complex and real equations: If one can check the complex equation, then he/she also gets the real equation by changing the isomorphism $\mathbb{R}^3 \cong \mathbb{R} \times \mathbb{C}$.

As is well-known, a monopole (A, Φ) on $\mathbb{R}^3 = \{(x_1, x_2, x_3)\}$ can be identified with an $\mathbb{R}$-invariant instanton B on $\mathbb{R}^4 = \{(x_0, x_1, x_2, x_3)\}$. The operators $D_{A,t}$, $D_{A,t}^*$ correspond to the Dirac operators $D_{B,t}^+$, $D_{B,t}^-$ respectively, where the subscript t means that the operators are twisted by a flat connection $itdx_0$. Using the isomorphism $\mathbb{R}^4 \cong \mathbb{C}^2$, we have isomorphisms

$$S^+ = \Lambda^{0,0} \oplus \Lambda^{0,2}, \qquad S^- = \Lambda^{0,1},$$

and the Dirac operators are written as

$$D_{B,t}^+ = \sqrt{2}(\overline{\partial}_{B,t}, \overline{\partial}_{B,t}^*) \colon \Omega^{0,0}(E) \oplus \Omega^{0,2}(E) \to \Omega^{0,1}(E),$$
$$D_{B,t}^- = \sqrt{2}\begin{pmatrix} \overline{\partial}_{B,t}^* \\ \overline{\partial}_{B,t} \end{pmatrix} \colon \Omega^{0,1}(E) \to \Omega^{0,0}(E) \oplus \Omega^{0,2}(E).$$

Correspondingly, we denote components of $D_{A,t}$, $D_{A,t}^*$ by Dolbeault operators:

$$D_{A,t} = \sqrt{2}(\overline{\partial}_{A,t}, \overline{\partial}_{A,t}^*), \qquad D_{A,t}^* = \sqrt{2}\begin{pmatrix} \overline{\partial}_{A,t}^* \\ \overline{\partial}_{A,t} \end{pmatrix}.$$

Then the followings are "key identities" in our calculation:

$$[\overline{\partial}_{A,t}, \frac{\partial}{\partial t} - x_1] = 0, \qquad [\overline{\partial}_{A,t}, x_2 + ix_3] = 0, \tag{2.8}$$

where x_α is the multiplication of a coordinate function. These identities means that "$z_1 = -i\frac{\partial}{\partial t} + ix_1$ and $z_2 = x_2 + ix_3$ are holomorphic" which is true on $\mathbb{C}^2 = \{(z_1, z_2)\}$. We shall use this funny notation, hoping this causes no confusion.

In this setting, the formula (2.1) is

$$(\overline{\partial}_{A,t}, \overline{\partial}^*_{A,t}) \begin{pmatrix} \overline{\partial}^*_{A,t} \\ \overline{\partial}_{A,t} \end{pmatrix} = \begin{pmatrix} \overline{\partial}_{A,t}\overline{\partial}^*_{A,t} & \overline{\partial}^*_{A,t}\overline{\partial}^*_{A,t} \\ \overline{\partial}_{A,t}\overline{\partial}_{A,t} & \overline{\partial}^*_{A,t}\overline{\partial}_{A,t} \end{pmatrix} = \frac{1}{2}\begin{pmatrix} \Delta_{A,t} & 0 \\ 0 & \Delta_{A,t} \end{pmatrix}, \tag{2.9}$$

where $\Delta_{A,t} = \nabla_A^* \nabla_A - (\Phi - it)^2$.

Proof of Proposition 2.7. Let $G_{A,t}$ denote the Green's operator $\Delta_{A,t}^{-1}$. Then the orthogonal projection π is given by

$$\pi = 1 - D_{A,t}(1_S \otimes G_{A,t})D^*_{A,t}.$$

Let $\psi \in V_t$, i.e. an L^2-solution of $D^*_{A,t}\psi = 0$. By the definitions of ∇ and T_α

$$\begin{aligned} iz_1\psi - (\frac{d}{dt} + 2\alpha)\psi &= D_{A,t}(1_S \otimes G_{A,t})D^*_{A,t}(iz_1\psi) \\ iz_2\psi - 2\beta\psi &= D_{A,t}(1_S \otimes G_{A,t})D^*_{A,t}(iz_2\psi). \end{aligned} \tag{2.10}$$

Using the "Dolbeault" operators and the formula (2.8), (2.9), we find

$$iz_2(\frac{d}{dt} + 2\alpha)\psi - 2iz_1\beta\psi = 2\overline{\partial}_{A,t}\{iz_1 G_{A,t}\overline{\partial}^*_{A,t}(iz_2\psi) - iz_2 G_{A,t}\overline{\partial}^*_{A,t}(iz_1\psi)\}. \tag{2.11}$$

Projecting to V_t, we get the complex equation

$$\frac{d\beta}{dt} + 2[\alpha, \beta] = 0.$$

This completes the proof. □

3. From Nahm's equations to Monopoles

In this section, we shall construct an SU(2)-monopole from a solution of the Nahm's equations.

Suppose that we are given Nahm data satisfying the conditions B in Sect. 1. Let consider the Sobolev space $\mathbb{C}^2 \otimes W_0^{1,2}(I;V)$ of sections of $\mathbb{C}^2 \otimes V$ whose derivatives are in L^2 and the boundary values are 0. Similarly let $\mathbb{C}^2 \otimes L^2(I;V)$ be the space of L^2 sections. For each $x \in \mathbb{R}^3$, define an operator $\mathfrak{D}_x: \mathbb{C}^2 \otimes W_0^{1,2}(I;V) \to \mathbb{C}^2 \otimes L^2(I;V)$ by

$$\mathfrak{D}_x = 1_{\mathbb{C}^2} \otimes \nabla_t + \sum_{\alpha=1}^{3}(e_\alpha \otimes T_\alpha - ix_\alpha e_\alpha \otimes 1_V),$$

where $\{e_1, e_2, e_3\}$ is the standard basis for $\mathfrak{su}(2)$ (see Sect. 1). In the matrix notation, this is equal to

$$\mathfrak{D}_x = \begin{pmatrix} \frac{d}{dt} + 2\alpha & 2\beta^* \\ 2\beta & \frac{d}{dt} - 2\alpha^* \end{pmatrix} - \begin{pmatrix} -x_1 & -i\overline{z}_2 \\ iz_2 & x_1 \end{pmatrix},$$

where $z_2 = x_2 + ix_3$ as before. Let $\mathfrak{D}^*_x$ be the formal adjoint operator of $\mathfrak{D}_x$, which is given by

$$\mathfrak{D}^*_x = \begin{pmatrix} -\frac{d}{dt} + 2\alpha^* & 2\beta^* \\ 2\beta & -\frac{d}{dt} - 2\alpha \end{pmatrix} - \begin{pmatrix} -x_1 & -i\overline{z}_2 \\ iz_2 & x_1 \end{pmatrix}.$$

The Nahm's equations imply

$$\mathfrak{D}_x^*\mathfrak{D}_x = 1_{\mathbb{C}^2} \otimes \left(\nabla_t^* \nabla_t + \sum_{\alpha=1}^{3} (T_\alpha - i x_\alpha)^* (T_\alpha - i x_\alpha) \right). \tag{3.1}$$

This identity is an analogue of (2.1). Then one can show that $\operatorname{Ker} \mathfrak{D}_x = 0$ for all $x \in \mathbb{R}^3$, so $\operatorname{Ker} \mathfrak{D}_x^*$ forms a vector bundle E over $\mathbb{R}^3$. The index is equal to -2 [Hi3], so $\operatorname{rank} E = 2$. Since E is a subbundle of the trivial bundle $\underline{\mathbb{C}^2 \otimes L^2(I;V)}$ over $\mathbb{R}^3$, it inherits a hermitian metric and a connection A. More precisely, if p is the projection onto E,

$$d_A = p \circ d.$$

We define the Higgs field Φ by

$$\Phi = p \circ it.$$

Then Hitchin shows that

Theorem 3.2. *The connection A and Higgs field Φ satisfy the Bogomolny equation $*R_A = d_A\Phi$ and the boundary condition A2.*

Proof. We shall give the proof for the Bogomolny equation. Our proof is "dual" to that of Proposition 2.7. For the proof of the boundary condition, see [Hi3].

Let define

$$\sigma_x = \begin{pmatrix} \frac{d}{dt} + 2\alpha + x_1 \\ 2\beta - i z_2 \end{pmatrix}, \qquad \tau_x = (2\beta - i z_2, -\frac{d}{dt} - 2\alpha - x_1).$$

Let F_x be the inverse of $\nabla_t^*\nabla_t + \sum_{\alpha=1}^3 (T_\alpha - i x_\alpha)^*(T_\alpha - i x_\alpha)$. Then the orthogonal projection p is given by

$$p = 1 - \mathfrak{D}_x(1_{\mathbb{C}^2} \otimes F_x)\mathfrak{D}_x^* = 1 - \sigma_x F_x \sigma_x^* - \tau_x^* F_x \tau_x.$$

Consider the following operators (cf. Sect. 2)

$$\mathbb{D} = D + it \colon \Gamma(S \otimes \underline{\mathbb{C}^2 \otimes L^2(I;V)}) \to \Gamma(S \otimes \underline{\mathbb{C}^2 \otimes L^2(I;V)}),$$
$$\mathbb{D}^* = D - it \colon \Gamma(S \otimes \underline{\mathbb{C}^2 \otimes L^2(I;V)}) \to \Gamma(S \otimes \underline{\mathbb{C}^2 \otimes L^2(I;V)}),$$

where D is the Dirac operator associated with the trivial monopole on $\underline{\mathbb{C}^2 \otimes L^2(I;V)}$. Denote by $\overline{\partial}$, $\overline{\partial}^*$ the associated "Dolbeault" operators. Namely

$$\mathbb{D} = \sqrt{2}(\overline{\partial}, \overline{\partial}^*), \qquad \mathbb{D}^* = \sqrt{2}\begin{pmatrix} \overline{\partial}^* \\ \overline{\partial} \end{pmatrix}.$$

If we define $\overline{\partial}_{A,0}$ (we set the parameter $t = 0$.) from (A, Φ) as in Sect. 2, we find

$$\overline{\partial}_{A,0} = p\overline{\partial} = (1 - \sigma_x F_x \sigma_x^* - \tau_x^* F_x \tau_x)\overline{\partial}.$$

Then

$$\overline{\partial}_{A,0}\overline{\partial}_{A,0} = 0 \tag{3.3}$$

follows from $\overline{\partial}\,\overline{\partial} = 0$ and the identities

$$[\overline{\partial}, \sigma] = 0, \qquad [\overline{\partial}, \tau] = 0 \tag{3.4}$$

which mean that σ and τ are "holomorphic". These are analogue of (2.8). Now changing the complex structure, we get the full Bogomolny equation from (3.3). □

4. Completeness

We now study the composition of the transformations given in previous sections.

$$\begin{array}{ccccc} \text{a monopole} & \overset{\S 2}{\Longrightarrow} & \text{Nahm data} & \overset{\S 3}{\Longrightarrow} & \text{a new monopole} \\ (A, \Phi) & & (\nabla, T_\alpha) & & (A', \Phi') \end{array}$$

Starting from a monopole (A, Φ) with the monopole charge k, we construct Nahm data T_α satisfying the conditions B in Sect. 2. Then we can construct another monopole (A', Φ') from this data as in Sect. 3. The aim of this section is to show that these data (A, Φ) and (A', Φ') are gauge equivalent. This will show that all monopoles arise by the construction given in Sect. 3.

First we shall construct a bundle map from the original bundle E to a new bundle E' on which (A', Φ') lives. Fix $t \in I$. Let $\psi \in \mathbb{C}^2 \otimes V_t$. Since V_t is a subspace of $L^2(\mathbb{R}^3; S \otimes E)$, we can define a section of $S \otimes \mathbb{C}^2 \otimes E$ by

$$K_t \psi = G_{A,t}[D^*_{A,t}, (-\frac{d}{dt} + i\underline{x})]\, \psi, \tag{4.1}$$

where $\underline{x} = \sum_{\alpha=1}^3 x_\alpha e_\alpha$. The commutator $[D^*_{A,t}, (-\frac{d}{dt} + i\underline{x})]$ is given by a Clifford multiplication of a constant vector, so can be applied to ψ which is defined only at t. Moreover, since $\psi \in \operatorname{Ker} D^*_{A,t}$, we have

$$K_t \psi = G_{A,t} D^*_{A,t} (-\frac{d}{dt} + i\underline{x})\, \psi.$$

Using the identification of S with $\mathbb{C}^2$, we have a contraction map

$$\omega \colon S \otimes \mathbb{C}^2 \ni (s_1, s_2) \otimes (t_1, t_2) \mapsto s_1 t_2 - s_2 t_1 \in \mathbb{C}.$$

Then the map $V_t \ni \psi \mapsto (\omega K_t \psi)(x) \in E_x$ gives a bundle map from the trivial bundle $\mathbb{C}^2 \otimes \underline{V_t}$ over $\mathbb{R}^3$ to E. Taking the hermitian adjoint, and moving t, we finally obtain a bundle map

$$\kappa \colon E \to \mathbb{C}^2 \otimes \underline{\Gamma(I; V)}.$$

First we show that the image of κ is in $\mathbb{C}^2 \otimes W^{-1,2}(I; V)$. As is obtained in Sect. 2, we have an estimate

$$\|G_{A,t}\varphi\|_{L^6} \le C(1+t)^{-1} \|\varphi\|_{L^2} \qquad \text{for } \varphi \in L^2(S \otimes E),$$

if t is near -1. Since $\psi \in \mathbb{C}^2 \otimes V_t = \mathbb{C}^2 \otimes \operatorname{Ker} D^*_{A,t}$ satisfies an elliptic partial differential equation, the above estimate and the L^p-estimates (cf. [GT, Chapter 9]) give us

$$|(\omega K_t \psi)(x)| \le C(1+t)^{-1} \|\psi\|_{L^2}$$

for some constant C independent of t (which may depend on x). So if $f \in \mathbb{C}^2 \otimes W_0^{1,2}(I; V)$, we have

$$\left| \int_{-1}^{1} (\omega K_t f(t))(x)\, dt \right| \le C \|f\|_{C_0^{1/2}} \le C \|f\|_{W_0^{1,2}}.$$

This means that the image of κ is in the dual space of $W_0^{1,2}$, i.e. $W^{-1,2}$.

We will prove that the image of κ is, in fact, contained in L^2, later. So the following proposition merely means that the image of κ satisfies a certain differential equation at this moment, but later it will mean that κ is a bundle map from E to E'.

Proposition 4.2. *For each $x \in \mathbb{R}^3$ the image $\kappa(E_x)$ is contained in* Ker $\mathfrak{D}_x^*$.

Proof. The calculation is the straightforward adaptation of that in [KN, Proposition 6.1].

We rewrite (4.1) by using the complex notation as in Sect. 2. For $\psi = (\psi_1, \psi_2) \in \mathbb{C}^2 \otimes C_0^\infty(I; V_t)$ the section $\omega K_t \psi$ can be rewritten as

$$\sqrt{2}\, G_{A,t} \left\{ \overline{\partial}_{A,t}^* (i z_2 \psi_1 - i z_1 \psi_2) + \overline{\partial}_{A,t} (i\overline{z}_1 \psi_1 + i\overline{z}_2 \psi_2) \right\},$$

where z_1, $\overline{z}_1$ are as in Sect. 2. We have the following identities (cf. [KN, Lemma 6.2]):

$$(4.3) \qquad [\overline{\partial}_{A,t}\,, i\overline{z}_1] = [\overline{\partial}_{A,t}^*\,, iz_2], \qquad [\overline{\partial}_{A,t}^*\,, iz_1] = -[\overline{\partial}_{A,t}\,, i\overline{z}_2],$$

which can be checked easily. Hence we get

$$(4.4) \qquad \omega K_t \psi = 2\sqrt{2}\, G_{A,t} \overline{\partial}_{A,t}^* (iz_2\psi_1 - iz_1\psi_2) = 2\sqrt{2}\, G_{A,t}\overline{\partial}_{A,t}(i\overline{z}_1\psi_1 + i\overline{z}_2\psi_2).$$

From (2.11), we find

$$G_{A,t}\overline{\partial}_{A,t}^*\{iz_2(\frac{d}{dt} + 2\alpha)\psi_1 - 2iz_1\beta\psi_1\} = iz_1 G_{A,t}\overline{\partial}_{A,t}^*(iz_2\psi_1) - iz_2 G_{A,t}\overline{\partial}_{A,t}^*(iz_1\psi_1).$$

The integration of the right hand side is equal to

$$(4.5) \qquad \int_{-1}^{1} -x_1 G_{A,t}\overline{\partial}_{A,t}^*(iz_2\psi_1) - iz_2 G_{A,t}\overline{\partial}_{A,t}^*(iz_1\psi_1)\, dt,$$

where the term

$$\int_{-1}^{1} \frac{\partial}{\partial t} G_{A,t}\overline{\partial}_{A,t}^*(iz_2\psi_1)\, dt = \lim_{t\to 1} G_{A,t}\overline{\partial}_{A,t}^*(iz_2\psi_1) - \lim_{t\to -1} G_{A,t}\overline{\partial}_{A,t}^*(iz_2\psi_1)$$

drops because ψ_1 vanishes near the boundary. (Remember $iz_1 = \frac{\partial}{\partial t} - x_1$.) Similarly we have formula

$$(4.6) \qquad \begin{aligned} &\int_{-1}^{1} G_{A,t}\overline{\partial}_{A,t}\{2i\overline{z}_1\beta^*\psi_2 + i\overline{z}_2(-\frac{d}{dt} + 2\alpha^*)\psi_2\}\, dt \\ &= \int_{-1}^{1} -i\overline{z}_2 G_{A,t}\overline{\partial}_{A,t}(i\overline{z}_1\psi_2) + x_1 G_{A,t}\overline{\partial}_{A,t}(i\overline{z}_2\psi_2)\, dt. \end{aligned}$$

Then (4.5), (4.6) and (4.4) imply $\kappa^*\mathfrak{D}\psi = 0$. Since ψ is arbitrary, we have $\mathfrak{D}^*\kappa = 0$. □

Proposition 4.7. $\operatorname{Im}\kappa \subset \mathbb{C}^2 \otimes L^2(I; V)$.

Proof. Elements represented in the form

$$\mathfrak{D}_x f + g \qquad f \in \mathbb{C}^2 \otimes C_0^\infty(I; V),\ g \in \operatorname{Ker}\mathfrak{D}_x^* \cap L^2$$

are dense in $\mathbb{C}^2 \otimes L^2(I; V)$. The $\operatorname{Im}\kappa$ is orthogonal to $\mathfrak{D}_x f$ by Proposition 4.2. As used in [Hi3], an L^2-solution of $\mathfrak{D}_x^* g = 0$ is $O((1-t)^{(k-1)/2})$ near $t = 1$. Hence when $k > 1$, the L^2-inner product

$$\int_{-1}^{1} \langle g, \kappa(e)\rangle\, dt$$

is finite. When $k = 1$, T_α is bounded. So the equation

$$0 = \mathfrak{D}_x^*\kappa(e) = \left(-1_{\mathbb{C}^2} \otimes \nabla_t + \sum_{\alpha=1}^{3}(e_\alpha \otimes T_\alpha - ix_\alpha e_\alpha \otimes 1_V)\right)\kappa(e) \quad (e \in E_x)$$

implies that $\kappa(e)$ is bounded. So in either case, we have $\kappa(e) \in L^2$. □

Proposition 4.8. *The bundle map $\kappa: E \to E'$ respects the metric (up to constant), the connection and the Higgs field.*

Proof. Let define

$$L_x(g) = \int_{-1}^{1} (\omega K_t g)(x)\, dt \in E_x$$

for $g \in \mathbb{C}^2 \otimes L^2(I;V)$. Take a local section ψ of $E' = \operatorname{Ker} \mathfrak{D}^*$. We regard $L_x(\psi(y))$ as a local section of $p_1^*(E)$ over $\mathbb{R}^3 \times \mathbb{R}^3$. The section $\kappa^*(\psi)$ is obtained by restricting $L_x(\psi(y))$ to the diagonal $x = y$. Then we have

$$\overline{\partial}^x_{A,0}\kappa^*(\psi) = \overline{\partial}^x_{A,0} L_x(\psi(y))\Big|_{y=x} + \overline{\partial}^y_{A,0} L_x(\psi(y))\Big|_{y=x}, \tag{4.9}$$

where the superscript x or y for $\overline{\partial}_{A,0}$ indicates the variable with respect to the differentiation is done. Using (2.10), we have

$$\begin{aligned}
&2\overline{\partial}^x_{A,t} G^x_{A,t} \overline{\partial}^{x*}_{A,t}(iz_2\psi_1(y) - iz_1\psi_2(y))\Big|_{y=x} \\
&= (iz_2 - 2\beta)\psi_1(x) - (iz_1 - (\frac{d}{dt} + 2\alpha))\psi_2(x) = -\frac{d}{dt}\psi_2(x),
\end{aligned}$$

where we have used $\mathfrak{D}^*_x\psi(x) = 0$ in the latter equality. The integration of the right hand side over I vanishes because

$$\lim_{t\to\pm1} \frac{\psi_2(x)}{\|\psi_2\|_{L^2}} = 0.$$

(This follows from the study of the asymptotic behaviour as $t \to \pm1$ in Sect. 2.) Noticing $\overline{\partial}_{A,t} = \overline{\partial}_{A,0} - it$, $\Phi' = p \circ it$ and $\kappa^*\mathfrak{D} = 0$, we get

$$\begin{aligned}
\overline{\partial}^x_{A,0} L_x(\psi(y))\Big|_{y=x} &= 2\sqrt{2}\int_{-1}^{1} it\; G^x_{A,t}\overline{\partial}^{x*}_{A,t}(iz_2\psi_1(y) - iz_1\psi_2(y))\Big|_{y=x} dt \\
&= 2\sqrt{2}\int_{-1}^{1} G^x_{A,t}\overline{\partial}^{x*}_{A,t}(iz_2\Phi'(y)\psi_1(y) - iz_1\Phi'(y)\psi_2(y))\Big|_{y=x} dt.
\end{aligned}$$

Substituting into (4.9) and using $\kappa^*\mathfrak{D} = 0$ again, we obtain

$$\overline{\partial}_{A,0}\kappa^*(\psi) = \kappa^*(\overline{\partial}_{A',0}\psi).$$

Changing the isomorphism $\mathbb{R}^3 \cong \mathbb{R} \times \mathbb{C}$, we can conclude that κ respects the connection and the Higgs field. If κ is a zero map, $\omega K_t\psi = 0$ for all $\psi \in V_t$. But the equation implies $\psi = 0$, and it is a contradition. Since monopole connections are irreducible, κ preserves the fiber metrics up to a constant factor. □

5. Uniqueness

We now finish the proof of our main theorem.

$$\begin{matrix} \text{Nahm data} & \overset{\S3}{\Longrightarrow} & \text{a monopole} & \overset{\S2}{\Longrightarrow} & \text{new Nahm data} \\ (\nabla, T_\alpha) & & (A, \Phi) & & (\nabla', T'_\alpha) \end{matrix}$$

Starting from Nahm data V, ∇, T_α ($\alpha = 1,2,3$), we construct a rank 2 vector bundle E over $\mathbb{R}^3$ with a connection A and a Higgs field Φ which satisfy the Bogomolny equation in Sect. 3. We then get new Nahm data V', ∇', T'_α by the transform in Sect. 2. The aim of this section is to show that there exists a isomorphism $V \cong V'$ under which ∇ and T_α correspond to ∇' and T'_α. This shows the uniqueness of the Nahm data corresponding to a monopole.

The proof is exactly "dual" to that of the completeness. Fix $x \in \mathbb{R}^3$. Let $f \in S \otimes E_x$. Since E_x is a subspace of $\mathbb{C}^2 \otimes L^2(I;V)$, we can define a section of $S \otimes \mathbb{C}^2 \otimes V$ by

$$F_x\,[\mathfrak{D}_x^*, \mathbb{D}^*]\, f,$$

where $\mathbb{D}^*$ is the operator acting on sections of the bundle $\underline{S \otimes \mathbb{C}^2 \otimes L^2(I;V)}$ over $\mathbb{R}^3$ defined in Sect. 3. Although f is defined only at x, the commutator $[\mathfrak{D}_x^*, \mathbb{D}^*]$ is equal to multiplication by a constant vector, so can be applied to f. Contracting the $S \otimes \mathbb{C}^2$-factor by ω, taking the hermitian adjoint and moving $x \in \mathbb{R}^3$, we obtain a bundle map

$$\lambda\colon V \to \underline{\Gamma(\mathbb{R}^3; S \otimes E)}$$

over I.

First we show that the image of λ is in $L^{2+\mu}$ for any $\mu > 0$. We use the notation C to denote a general constant; C may be different in different equations. If $g \in W_0^{1,2}(I;V)$, we have an estimate

$$\begin{aligned}
&(\nabla_t^*\nabla_t g + \sum_{\alpha=1}^{3}(T_\alpha - ix_\alpha)^*(T_\alpha - ix_\alpha)g, g)_{L^2(I;V)} \\
&= \|\nabla_t g\|^2_{L^2(I;V)} + \sum_{\alpha=1}^{3}\|(T_\alpha - ix_\alpha)g\|^2_{L^2(I;V)} \qquad (5.1)\\
&\geq \|\nabla_t g\|^2_{L^2(I;V)} - \delta\sum_{\alpha=1}^{3}\|T_\alpha g\|^2_{L^2(I;V)} + \frac{r^2\delta}{2}\|g\|^2_{L^2(I;V)},
\end{aligned}$$

where $\delta < 1$ is a positive number, which will be fixed later. Using the Sobolev inequlity

$$\|g\|_{C^{1/2}(I;V)} \leq C\|\nabla_t g\|_{L^2(I;V)}$$

and the asymptotic behaivour of T_α as $t \to \pm 1$, we find

$$\|T_\alpha g\|^2_{L^2(I;V)} \leq C\|\nabla_t g\|^2_{L^2(I;V)}.$$

Substituting this inequality into (5.1) and choosing δ sufficiently small, we get

$$(\nabla_t^*\nabla_t g + \sum_{\alpha=1}^{3}(T_\alpha - ix_\alpha)^*(T_\alpha - ix_\alpha)g, g)_{L^2(I;V)} \geq \frac{1}{C}(\|\nabla_t g\|^2_{L^2(I;V)} + r^2\|g\|^2_{L^2(I;V)}).$$

Hence for $g = F_x h$, we have

$$\|F_x h\|_{C^{1/2}(I;V)} \leq \frac{C}{r}\|h\|_{L^2(I;V)}.$$

Fix an $\varepsilon > 0$ and take $f \in L^{2-\varepsilon}(\mathbb{R}^3; S \otimes E)$. Then for $v \in V_t$, we have

$$\begin{aligned}
\int_{\mathbb{R}^3}\langle\lambda(v), f(x)\rangle\, dx &= \int_{\mathbb{R}^3}\langle v, (\omega F_x\,[\mathfrak{D}_x^*, \mathbb{D}^*]\, f(x))(t)\rangle\, dx \\
&\leq C\|v\|_{V_t}\int_{\mathbb{R}^3}(1+r)^{-1}\|f(x)\|_{S\otimes E_x}\, dx \leq C_\varepsilon\|v\|_{V_t}\,\|f\|_{L^{2-\varepsilon}(\mathbb{R}^3; S\otimes E)},
\end{aligned}$$

where C_ε is a constant depending on ε. This shows that $\lambda(v) \in L^{2+\mu}$ for any $\mu > 0$.

Proposition 5.2. *For each $t \in I$ the image $\lambda(V_t)$ is contained in* Ker $D^*_{A,t}$.

The proof is exactly "dual" to that of (4.2), and we skip it. Once we obtain the above, we deduce the following since any $L^{2+\mu}$-solution of $D^*_{A,t}\varphi = 0$ decays exponentially.

Corollary 5.3. *λ defines a bundle map from V to V'.*

Finally we get the following which can be proved by the argument similar to (4.8).

Proposition 5.4. *The bundle map $\lambda: V \to V'$ intertwines the Nahm data.*

6. Metrics on moduli spaces

Our transform identifies the (framed) moduli space of SU(2)-monopoles of charge k with the moduli space of the solutions of Nahm's equations of rank k. These moduli spaces are well-known to admit hyper-Kähler metrics. Atiyah and Hitchin conjectured that our transform is actually a hyper-Kähler isometry [AH, p.126]. We shall verify this conjecture. The corresponding results for the Fourier transforms of instantons on 4-tori and on ALE spaces are proved respectively in [BB] and [KN] (in the case of $\mathbb{R}^4$ independently in [Ma]).

We shall review the construction of a hyper-Kähler structure on the monopole moduli space very quickly. See [AH] and the reference therein for detail. (The Analytical footing was established by Taubes [Ta].)

We introduce an equivalence relation $\sim$ on the space of SU(2)-monopoles of charge k by defining $(A, \Phi) \sim (A', \Phi')$ if and only if (A, Φ) and (A', Φ') are gauge-equivalent under a gauge transformation converging to the identity as $x \to \infty$. Let denote $\mathfrak{M}_k$ the set of equivalence classes. Then the following is well-known:

Proposition 6.1. *The space $\mathfrak{M}_k$ has a structure of a smooth manifold and its tangent space at $[(A, \Phi)]$ is identified with the space of (a, ϕ) which are in L^2 and satisfy the equations*

$$* d_A a - d_A \phi + [\Phi, a] = 0$$
$$* d_A * a - [\Phi, \phi] = 0.$$

Here a and ϕ are an $\mathfrak{su}(2)$-valued 1-form and function respectively.

The second equation is a linearization of Bogomolny equations, while the first one means that (a, ϕ) is orthogonal to the orbit of gauge group action.

The space of all pairs (a, ϕ) has a structure of quaternion module. In fact, if $a = a_1 dx_1 + a_2 dx_2 + a_3 dx_3$, then (a, ϕ) corresponds to the $\mathfrak{su}(2) \otimes \mathbb{H}$-valued function $\phi + a_1 I + a_2 J + a_3 K$, where I, J, K are the usual basis of imaginary quaternions. The equations in (6.1) are $\mathbb{H}$-invariant.

The L^2-inner product induces a Riemannian metric on $\mathfrak{M}_k$. Then one can show that

Proposition 6.2. *The almost complex structures I, J, K are parallel with respect to the Levi-Civita connection of the Riemannian metric. Hence $\mathfrak{M}_k$ has a structure of hyper-Kähler manifold.*

In fact, the Bogomolny equation can be viewed as a hyper-Kähler moment map (see [HKLR]) associated with the action of the gauge group, and the moduli space is a hyper-Kähler quotient of an infinite dimensional quaternion module.

The construction of a hyper-Kähler structure on the moduli space of the solutions to Nahm's equations of rank k is similar to the above. We fix a trivialization of a bundle V so that the connection is given by $\nabla = \frac{d}{dt} + T_0$ where T_0 is a skew-adjoint endomorphism. We say two Nahm data (T_α), (T'_α) $(\alpha = 0,1,2,3)$ are equivalent if they are gauge equivalent under a gauge transformation converging to the identity at the end points of the interval. We denote by $\mathfrak{N}_k$ the set of equivalence classes. Then

Proposition 6.3. *The space $\mathfrak{N}_k$ has a structure of a smooth manifold and its tangent space at $[(T_\alpha)]$ is identified with the space of (t_0, t_1, t_2, t_3) which are in L^2 and satisfy the equations*

$$\frac{dt_0}{dt} + [T_0, t_0] + [T_1, t_1] + [T_2, t_2] + [T_3, t_3] = 0$$

$$\frac{dt_\alpha}{dt} + [T_0, t_\alpha] - [T_\alpha, t_0] + \sum_{\beta,\gamma=1}^{3} \varepsilon_{\alpha\beta\gamma}[T_\beta, t_\gamma] = 0, \qquad \alpha = 1,2,3.$$

The metric on $\mathfrak{N}_k$ is defined by the L^2-inner product. We give an $\mathbb{H}$-module structure to the tangent space by identifying (t_0, t_1, t_2, t_3) with $t_0 + t_1 I + t_2 J + t_3 K$.

Proposition 6.4. *The manifold $\mathfrak{N}_k$ together with the above structures is hyper-Kähler.*

Now our main result in this section is

Theorem 6.5. *The transform $\Xi\colon \mathfrak{M}_k \to \mathfrak{N}_k$ given in Sect. 2 is a hyper-Kähler isometry up to a constant factor.*

The proof of Theorem 6.5 is very similar to that for instantons on ALE spaces [KN].

Suppose that a family (T^s_α) $(-\varepsilon < s < \varepsilon)$ of solutions of Nahm's equations is given. For brevity, we omit the superscript s. Let e_μ be a unitary frame field for $E = \operatorname{Ker}\mathfrak{D}^*$. Then the derivative δe_μ with respect s satisfies $\mathfrak{D}^*\delta e_\mu = -(\delta\mathfrak{D}^*)e_\mu$. If we normalize δe_μ by requiring $\delta e_\mu \perp E$, this equation implies

$$\delta e_\mu = -\mathfrak{D}(1_{\mathbf{C}^2} \otimes F)(\delta\mathfrak{D}^*)e_\mu.$$

The derivative of the connection $A_{\mu\nu} = \langle de_\mu, e_\nu\rangle$ and the Higgs field $\Phi_{\mu\nu} = \langle ite_\mu, e_\nu\rangle$ are given by

$$\begin{aligned} \delta A_{\mu\nu} &= \langle(\delta\mathfrak{D}^*)e_\mu, (1_{\mathbf{C}^2} \otimes F)\mathfrak{D}^* de_\nu\rangle - \langle(1_{\mathbf{C}^2} \otimes F)\mathfrak{D}^* de_\mu, (\delta\mathfrak{D}^*)e_\nu\rangle \\ \delta\Phi_{\mu\nu} &= \langle(\delta\mathfrak{D}^*)e_\mu, (1_{\mathbf{C}^2} \otimes F)\mathfrak{D}^* ite_\nu\rangle - \langle(1_{\mathbf{C}^2} \otimes F)\mathfrak{D}^* ite_\mu, (\delta\mathfrak{D}^*)e_\nu\rangle. \end{aligned} \tag{6.6}$$

Next suppose that a family (A^s, Φ^s) of monopoles is given. We omit the superscript s. Let v_i be a unitary frame field for $V = \bigcup_t \operatorname{Ker} D^*_{A,t}$. Then the derivative of $(T_\alpha)_{ij} = \langle ix_\alpha v_i, v_j\rangle$ and of $(T_0)_{ij} = \langle(\nabla_t - \frac{d}{dt})v_i, v_j\rangle$ are given by

$$\begin{aligned} \delta(T_\alpha)_{ij} &= \langle(\delta A)\cdot v_i - (\delta\Phi)v_i\,,\, (1_S \otimes G_{A,t})D^*_{A,t}(ix_\alpha v_j)\rangle \\ &\qquad - \langle(1_S \otimes G_{A,t})D^*_{A,t}(ix_\alpha v_i)\,,\, (\delta A)\cdot v_j - (\delta\Phi)v_j\rangle \\ \delta(T_0)_{ij} &= \langle(\delta A)\cdot v_i - (\delta\Phi)v_i\,,\, (1_S \otimes G_{A,t})D^*_{A,t}\frac{dv_j}{dt}\rangle \\ &\qquad - \langle(1_S \otimes G_{A,t})D^*_{A,t}\frac{dv_i}{dt}\,,\, (\delta A)\cdot v_j - (\delta\Phi)v_j\rangle, \end{aligned} \tag{6.7}$$

where $(\delta A)\cdot v_i = \sum_{\alpha=1}^3 \delta A(\frac{\partial}{\partial x_\alpha})\frac{\partial}{\partial x_\alpha}\cdot v_i$.

We fix a particular complex structure I, and regard the moduli spaces $\mathfrak{M}_k$ and $\mathfrak{N}_k$ as Kähler manifolds. First we shall show that the differential $d\Xi$ of our transformation respects the almost complex structures, i.e., Ξ is a holomorphic map. Then changing the complex structure, one can show that Ξ is holomorphic with respect to each complex structure I, J, K.

We rewrite (6.6) in the complex notation. We identify the monopole (A, Φ) with an invariant instanton on $\mathbb{R}^4 \cong \mathbb{C}^2$, hence $(\delta A, \delta\Phi)$ can be considered as a 1-form on $\mathbb{C}^2$. Then $(0,1)$-part of (6.6) (up to a constant factor) is given by

$$\begin{aligned}&\langle(\delta\mathfrak{D}^*)e_\mu, (1_{\mathbf{C}^2}\otimes F)\,\mathfrak{D}^*\overline{\partial}^*(e_\nu\omega_{\mathbf{C}})\rangle + \langle(1_{\mathbf{C}^2}\otimes F)\mathfrak{D}^*\overline{\partial}e_\mu, (\delta\mathfrak{D}^*)e_\nu\rangle \\ &= \langle(\delta\tau)e_\mu, F\tau\overline{\partial}^*(e_\nu\omega_{\mathbf{C}})\rangle + \langle F\sigma^*\overline{\partial}e_\mu, (\delta\sigma^*)e_\nu\rangle,\end{aligned} \tag{6.8}$$

where the inner product is taken over the fiber component, have nothing to do with the form component, and $\omega_{\mathbf{C}}$ is the $(0,2)$-form of unit length. Recall that $\{v_i\}$ be a unitary frame for V. Substituting $\lambda = (\omega\, F[\mathfrak{D}^*, \mathbb{D}^*])^*$, we find that (6.8) is equal to (up to a constant factor)

$$\langle(\delta\tau)e_\mu, v_i\rangle\langle\lambda(v_i), e_\nu\rangle + \langle v_i, (\delta\sigma^*)e_\nu\rangle\langle e_\mu, \varepsilon\lambda(v_i)\rangle. \tag{6.9}$$

Here ε is an endomorphism defined by

$$\Lambda^{0,1} \ni a d\overline{z}_1 + b d\overline{z}_2 \mapsto b d\overline{z}_1 - a d\overline{z}_2.$$

Similarly the $(0,1)$-component of (6.7) (up to a constant factor) is given by

$$\begin{aligned}&\langle(\delta A_2 + i\delta A_3, -\delta\Phi - i\delta A_1)\, v_i, e_\mu\rangle\langle\kappa(e_\mu), v_j\rangle \\ &\quad + \langle v_i, \varepsilon\kappa(e_\mu)\rangle\langle e_\mu, (-\delta\Phi + i\delta A_1, -\delta A_2 + i\delta A_3)v_j\rangle,\end{aligned} \tag{6.10}$$

where we take a unitary basis $\{\frac{1}{\sqrt{2}}d\overline{z}_1, \frac{1}{\sqrt{2}}d\overline{z}_2\}$ for $\Lambda^{0,1}$ and $\kappa(e_\mu) \in \mathbb{C}^2\otimes\overline{\Gamma(I;V)}$ is considered as a $(0,1)$-form. Here we have used the decomposition of the matrix $\delta A\cdot -\delta\Phi$: $S\otimes E \to S\otimes E$:

$$\begin{pmatrix} -\delta\Phi + i\delta A_1 & -\delta A_2 + i\delta A_3 \\ \delta A_2 + i\delta A_3 & -\delta\Phi - i\delta A_1 \end{pmatrix}.$$

If we apply I to (δT_α), $\delta\tau$ is multiplied by i and $\delta\sigma^*$ by $-i$. Hence the (6.8) is multiplied by i, and we verify the assertion. (REMARK that when we consider $(\delta A, \delta\Phi)$ as a tangent vector in the moduli space, its $(1,0)$-part is given by the $(0,1)$-part of $(\delta A, \delta\Phi)$, considered as 1-form.)

The only thing left to be proved is whether the map $d\Xi$ is isometry, i.e.

$$\langle d\Xi(\delta T_0, \delta T_1, \delta T_2, \delta T_3)\,,\, (\delta A, \delta\Phi)\rangle = c\langle(\delta T_0, \delta T_1, \delta T_2, \delta T_3)\,,\, d\Xi^{-1}(\delta A, \delta\Phi)\rangle. \tag{6.11}$$

holds for some constant c. This can be checked by using (6.9) and (6.10) and the fact : λ and κ are isometries up to constant factors. Rigorously speaking, we have not proved that the constants, up to which λ and κ are isometries, are independent of (A, Φ), T_α. So the constant c in (6.11) may change if we move (A, Φ). But we already observed that Ξ respects almost complex structures I, J, K. If a map between hyper-Kähler manifolds respects almost complex structures, it also respects the Levi-Civita connection. Hence c is a constant function.

7. Remark

A Fourier transform of invariant instantons. We explain, briefly and without proofs, how the transformation in Sects. 2 and 3 can be generalized for anti-self-dual connections on $\mathbb{R}^4$, invariant under a subgroup of translation $\Lambda \subset \mathbb{R}^4$. This is already noticed in [BB, p. 272], but it is worth while explaining again.

Let $(\mathbb{R}^4)^*$ denote the dual space of $\mathbb{R}^4$ and define

$$\Lambda^* = \{\lambda^* \in (\mathbb{R}^4)^* \mid \lambda^*(\lambda) \in \mathbb{Z}, \forall \lambda \in \Lambda\}.$$

For example, when $\Lambda = \mathbb{R}$ (this is our case), $\Lambda^* = \mathbb{R}^3$. Define a connection 1-form $\mathbb{A}$ on the trivial line bundle $\mathbb{L} \to \mathbb{R}^4 \times (\mathbb{R}^4)^*$ by

$$\mathbb{A} = -2\pi i \sum_{\alpha=0}^{3} q_\alpha dx_\alpha,$$

where x_α and q_α are dual linear coordinates on $\mathbb{R}^4$ and $(\mathbb{R}^4)^*$. The action of $\Lambda \times \Lambda^*$ on $\mathbb{R}^4 \times (\mathbb{R}^4)^*$ lifts to that on $\mathbb{L}$ by

$$\mathbb{L} = \mathbb{R}^4 \times (\mathbb{R}^4)^* \times \mathbb{C} \ni (x, q, \zeta) \mapsto (x + \lambda, q + \lambda^*, e^{2\pi i \lambda^*(x)}\zeta) \quad \text{for } (\lambda, \lambda^*) \in \Lambda \times \Lambda^*.$$

This action preserves $\mathbb{A}$.

Now suppose that we have a connection A on a bundle E over $\mathbb{R}^4$ invariant under Λ. For each $q \in (\mathbb{R}^4)^*$, consider the Dirac operator twisted by the connection $\mathbb{A}$ and A:

$$D^{\pm}_{A,q}: \Gamma(S^{\pm}_{\mathbb{R}^4} \otimes E \otimes \mathbb{L}|_{\mathbb{R}^4 \times \{q\}}) \to \Gamma(S^{\mp}_{\mathbb{R}^4} \otimes E \otimes \mathbb{L}|_{\mathbb{R}^4 \times \{q\}}),$$

where $S^{\pm}_{\mathbb{R}^4}$ is the spinor bundle over $\mathbb{R}^4$. Define

$$\hat{E}_q = \Lambda\text{-invariant part of the } L^2\text{-kernel of } D^{-}_{A,q},$$

where the L^2-metric is taken over $\mathbb{R}^4/\Lambda$. Assume that

a) Λ-invariant part of the L^2-kernel of $D^{+}_{A,q} = 0$,

b) $\hat{E} = \bigcup_q \hat{E}_q$ forms a vector bundle over $(\mathbb{R}^4)^*$.

Then considering $\hat{E}$ as a subbundle of a (may be infinite rank) vector bundle

$$\mathcal{H} = \bigcup_q \Lambda\text{-invariant part of } L^2(S^{-}_{\mathbb{R}^4} \otimes E \otimes \mathbb{L}|_{\mathbb{R}^4 \times \{q\}}),$$

we induce a metric and a connection $\hat{A}$ on $\hat{E}$. Here the connection on $\mathcal{H}$ is defined from A and $\mathbb{A}$. The action of $\Lambda \times \Lambda^*$ on $\mathbb{L}$ naturally induces an action of Λ^* on $\hat{E}$ and $\hat{A}$ is invariant under this action. Changing the role of x and q and using the dual connection $\mathbb{A}^*$ instead of $\mathbb{A}$, we define a similar transform (denoted by $\check{}$) from a Λ^*-invariant connection satisfying

a') Λ^*-invariant part of the L^2-kernel of $D^{+}_{A,x} = 0$,

b') $\check{E} = \bigcup_x \check{E}_x$ forms a vector bundle over $\mathbb{R}^4$.

to a Λ-invariant connection. Then one has

Theorem 7.1. *If A is anti-self-dual and satisfies a), b), then $\hat{A}$ is anti-self-dual and satisfies a'), b'). Moreover $\check{\hat{A}}$ is gauge equivalent to A.*

This theorem is not proven in full generality. In fact, we must put the condition on the asymptotic behaviour of the connection in order to ensure that the Fredholm theory is valid. In some cases, the connection is not defined over the whole space and may have singularities (as is observed in this paper). Such a modification and the precise proof are given only in the cases of $\Lambda = 0$ (ordinary instantons on $\mathbb{R}^4$ [ADHM, CG, DK]), $\Lambda = \mathbb{R}$ (monopoles on $\mathbb{R}^3$) and $\Lambda \cong \mathbb{Z}^4$ (instantons on torus $\mathbb{R}^4/\Lambda$ [BB, Sc, DK]).

Appendix

In this appendix, we shall give the proof of Lemma 2.6. The following proof is due to Toshiyuki Kobayashi.

Proof. Let denote by V_k the unique $(k+1)$-dimensional irreducible representation of SU(2). By the theorem of Peter-Weyl the space $L^2(\mathbb{P}^1; H^k)$ of L^2-sections of H^k (by which we denote the k-times tensor product of the hyperplane bundle) over $\mathbb{P}^1$ decomposes into

$$L^2(\mathbb{P}^1; H^k) = \bigoplus_{l \geq 0} V_{k+2l},$$

and the space $H^0(\mathbb{P}^1; \mathcal{O}(k))$ of holomorphic sections is the component V_k. The set of coordinate functions $\{x_1, x_2, x_3\}$ induces the 3-dimensional representation V_2. Then the multiplication of the coordinate function gives an SU(2)-equivariant map

$$m: L^2(\mathbb{P}^1; H^k) \otimes V_2 \to L^2(\mathbb{P}^1; H^k).$$

By the Clebsch-Gordan rule

$$V_k \otimes V_2 \cong V_{k+2} \oplus V_k \oplus V_{k-2}. \tag{A.1}$$

The map defined in Lemma 2.6 is the composition of

$$V_k \otimes V_2 \xrightarrow{\text{inclusion}\otimes\text{id}} L^2(\mathbb{P}^1; H^k) \otimes V_2 \xrightarrow{m} L^2(\mathbb{P}^1; H^k) \xrightarrow{\text{projection}} V_k. \tag{A.2}$$

It is SU(2)-equivariant, and must be a constant multiple of the projection onto the second component in (A.1).

On the other hand, the adjoint representation of SU(2) is also V_2. So we have a linear map

$$V_k \otimes V_2 \to V_k; \quad v \otimes X \mapsto Xv$$

where V_2 is regarded as(the complexification of) the Lie algebra $\mathfrak{su}(2)$ and it acts on V_k by the differential of the action of SU(2) on V_k. This map is also SU(2)-equivariant, so must be a constant multiple of the projection onto the second component in (A.1).

Finally we must check that the map (A.2) is non-zero. It is sufficient to show that the map

$$V_k \otimes V_2 \xrightarrow{\text{inclusion}\otimes\text{id}} L^2(\mathbb{P}^1; H^k) \otimes V_2 \xrightarrow{m} L^2(\mathbb{P}^1; H^k) = \bigoplus_{l \geq 0} V_{k+2l}$$

has rank strictly greater that $\dim V_{k+2}$. (Since it is SU(2)-equivariant, the image is contained in $V_k \oplus V_{k+2}$.) Elements in V_k can be represented by homogeneous polynomial in z_0 and z_1 of degree k:

$$z_0^k, z_0^{k-1} z_1, \ldots, z_0 z_1^{k-1}, z_1^k.$$

It is easy to see that if we multiply the above functions by $x_1 = \frac{2\mathrm{Re} z_0}{1+|z_0|^2}$ and $x_2 = \frac{2\mathrm{Im} z_0}{1+|z_0|^2}$ (where we take an affine coordinate $[z_0 : 1] \in \mathbb{P}^1$ by setting $z_1 = 1$), we obtain $2(k+1)$ linearly independent functions. Hence if $2(k+1) > k+3$, we are done. And in the exceptional case $k = 0, 1$, we can check the assertion case by case.

Acknowledgement. I would like to thank Shigetoshi Bando for a number of interesting discussions and for pointing out several errors in an earlier version, and Toshiyuki Kobayashi for informing me of the proof of Lemma 2.6.

REFERENCES

[ADHM] M.F. Atiyah, V. Drinfeld, N.J. Hitchin and Y.I. Manin, *Construction of instantons*, Phys. Lett. **65A** (1978), 185–187.

[AH] M.F. Atiyah and N.J. Hitchin, "Geometry and dynamics of magnetic monopoles," Princeton Univ. Press, Princeton, N.J., 1988.

[BB] P.J. Braam and P. van Baal, *Nahm's transformation for instantons*, Comm. Math. Phys. **122** (1989), 267–280.

[Ca] C. Callias, *Axial anomalies and index theorems on open spaces*, Comm. Math. Phys. **62** (1978), 213–234.

[CG] E. Corrigan and P. Goddard, *Construction of instantons and monopole solutions and reciprocity*, Ann. of Phys. **154** (1984), 253–279.

[Do] S.K. Donaldson, *Nahm's equations and the classification of monopoles*, Comm. Math. Phys. **96** (1984), 387–408.

[DK] S.K. Donaldson and P.B. Kronheimer, "The Geometry of Four-Manifolds," Oxford University Press, 1990.

[GT] D. Gilbarg and N.S. Trudinger, "Partial differential equations of second order, second edition," Springer, Berlin, Heidelberg, New York, 1983.

[Hi1] N.J. Hitchin, *Harmonic spinors*, Adv. Math **14** (1974), 1–55.

[Hi2] ___________, *Monopoles and Geodesics*, Comm. Math. Phys. **83** (1982), 579–602.

[Hi3] ___________, *On the construction of monopoles*, Comm. Math. Phys. **89** (1983), 145–190.

[HKLR] N.J. Hitchin, A. Karlhede, U. Lindström and M. Roček, *Hyperkähler metrics and supersymmetry*, Comm. Math. Phys. **108** (1987), 535–589.

[HM] J. Hurtubise and M.K. Murray, *On the construction of monopoles for the classical groups*, Comm. Math. Phys. **122** (1989), 35–89.

[JT] A. Jaffe and C.H. Taubes, "Vortices and monopoles," Birkhäuser, Boston, Basel, Stuttgart, 1980.

[KN] P.B. Kronheimer and H. Nakajima, *Yang-Mills instantons on ALE gravitational instantons*, Math. Ann. **288** (1990), 263–307.

[Ma] A. Maciocia, *Metrics on the moduli spaces of instantons over Euclidean 4-space*, Comm. Math. Phys. **135** (1991), 467–482.

[Na1] W. Nahm, *The construction of all self-dual multi-monopoles by the ADHM method*, in "Monopoles in quantum field theory," Craigie et al. (eds.), World Scientific, Singapore, 1982.

[Na2] ———, *Self-dual monopoles and calorons*, in "Lecture Notes in Physics," **201**, Springer, New York, 1984.

[Sc] H. Schenk, *On a generalised Fourier transform of instantons over flat tori*, Comm. Math. Phys. **116** (1988), 177–183.

[Ta] C.H. Taubes, *Stability in Yang-Mills theories*, Comm. Math. Phys. **91** (1983), 235–263.

15

Existence of Infinitely Many Solutions of a Conformally Invariant Elliptic Equation

SHOICHIRO TAKAKUWA
Department of Mathematics,
Tokyo Metropolitan University
Tokyo, Japan

1. INTRODUCTION

Let (M, g) be a compact connected Ricmannian manifold of dimension n (≥ 3). In this paper we consider the nonlinear eigenvalue problem

$$(1.1) \qquad L_g u := -\kappa \Delta_g u + R_g u = \lambda |u|^{N-2} u \qquad \text{in } M\,,$$
$$\kappa = \frac{4(n-1)}{n-2}\,, \qquad N = \frac{2n}{n-2}\,,$$

where Δ_g denotes the negative definite Laplacian and R_g is the scalar curvature of g. Equation (1.1) was first considered by Yamabe [Y]. He noticed that if a positive smooth function u satisfies (1.1) together with λ then the conformal metric $u^{N-2}g$ has constant scalar curvature λ. He observed that (1.1) is the Euler–Lagrange equation of the functional

$$Q(u) = E(u)/\|u\|_N^2 \qquad \text{for} \quad u \not\equiv 0\,,$$

where

$$E(u) = \int_M (\kappa |\nabla u|^2 + R_g u^2)\, dV \qquad \|u\|_N = \Big(\int_M |u|^N dV\Big)^{1/N}.$$

As shown in [LP], if u is a critical point of Q, then u satisfies (1.1) with $\lambda = E(u)/\|u\|_N^N$. Yamabe attempted to solve (1.1) by finding a function which attains the infimum

$$\lambda_g = \lambda_g(M) = \inf\{\, Q(u) \mid u \in C^\infty(M),\ u \not\equiv 0 \,\}\,.$$

More than twenty years after his pioneer work the existence of a minimizer was proved by Aubin [A1] and Schoen [Sc].

On non–minimizing critical points, however, only a few results have been proved. One of them is done by Ding [D] in case M is the standard sphere S^n. He proved the existence of infinitely many critical points of Q by using symmetry of the sphere.

The purpose of this paper is to extend Ding's result to more general manifolds. Our main result is the following.

Main Theorem. *Let (M, g) be a compact connected Riemannian manifold with positive scalar curvature. Suppose that a compact Lie group G of isometries satisfies*

(G1) *for any $x \in M$ the orbit $G(x) = \{ \gamma x \mid \gamma \in G \}$ is an infinite set.*

(G2) *the action of G is not transitive.*

Then, there exists a sequence $\{u_j\}_{j=0}^{\infty}$ of C^2 functions on M satisfying

(1) *each u_j is a critical point of Q, that is, u_j satisfies (1.1) with $\lambda = E(u_j)/\|u_j\|_N^N$.*

(2) *each u_j is invariant under the action of G,*

(3) $Q(u_j) \longrightarrow \infty$ *as* $j \to \infty$,

(4) *u_0 is a positive smooth function which attain the minimum of Q over G–invariant functions.*

In section 2 we prove the compactness lemma for G–invariant functions, which plays a crucial role in the proof of our main theorem. In case $M = S^n$ this lemma was proved by Ding [D] for a special class of G. Our proof differs from his. We make use of the concentration compactness lemma of P. L. Lions [Li] to prove the lemma for more general cases. In section 3 we prove our main theorem and discuss about some examples. In section 4 we state the similar result for the boundary value problem in Euclidean space.

2. Compactness Lemma

We denote by $H^1(M)$ the Sobolev space of L^2 functions whose derivatives are in $L^2(M)$. In this section we prove the following.

Lemma 2.1. *Let G be a group of isometries satisfying (G1). Then,the embedding of the subspace*

$$(2.1) \qquad X_G := \{\, u \in H^1(M) \mid u(\gamma x) = u(x) \qquad \text{for all } \gamma \in G\,, \quad \text{a.c. } x \in M \,\},$$

into $L^N(M)$ is compact.

To prove this lemma we need the following theorem.

Theorem 2.2. *Let $\{u_j\}$ be a sequence of $H^1(M)$ converges to some u weakly. Assume that $|u_j|^N dV$ converges to some measure ν weakly in the sense of measures. Then, there exist*

(1) *at most countable set $\mathcal{S} = \{x_i\}_{i\in I} \subset M$,*

(2) *a set $\{\nu_{x_i}\}_{i\in I} \subset \mathbb{R}_+$,*

such that

$$\nu = |u|^N dV + \sum_{i\in I} \nu_{x_i}\delta_{x_i}\,,$$

holds where δ_{x_i} is Dirac measure.

This theorem was proved by P. L. Lions in [Li]. For the proof of this theorem the reader is referred to Appendix.

Proof of Lemma 2.1. Take a bounded sequence $\{u_j\}$ of X_G. Taking a subsequence if necessary, we assume that

$$u_j \longrightarrow u \qquad \text{weakly in } H^1(M),$$
$$\qquad\qquad\qquad \text{a.e. on } M,$$

for some $u \in H^1(M)$. Since X_G is a closed subspace of $H^1(M)$, we get $u \in X_G$. From the Sobolev embedding theorem $\{u_j\}$ is bounded in $L^N(M)$. This means that the measure $|u_j|^N dV$ has the uniformly bounded total variation. Thus, we also assume that $|u_j|^N dV$ converges to some finite measure ν weakly. We use Theorem 2.2 to obtain

$$\nu = |u|^N dV + \sum_{i\in I} \nu_{x_i}\delta_{x_i}\,,$$

for some $\mathcal{S} = \{x_i\}_{i \in I}$ and $\{\nu_{x_i}\}_{i \in I}$. We first prove $\nu = |u|^N dV$. Assume that $\mathcal{S}$ is not empty. Since u_j and u lie in X_G, ν is also G–invariant. Hence, we have

$$\gamma x_i \in \mathcal{S}, \qquad \nu_{\gamma x_i} = \nu_{x_i} \qquad \text{for all } x_i \in \mathcal{S}, \text{ all } \gamma \in G.$$

Condition (G.1) implies that

$$\|\nu\| = \int_M d\nu \geq \sum_{i \in I} \nu_{x_i} = +\infty .$$

This contradicts to the fact that the measure ν is finite. Then, we obtain $\mathcal{S}$ is empty and

$$\lim_{j \to \infty} \|u_j\|_N = \|u\|_N .$$

From the theorem of Brezis–Lieb [BL], we obtain

$$\begin{aligned} \|u_j - u\|_N^N &= \|u_j\|_N^N - \|u\|_N^N + o(1), \\ &= o(1). \end{aligned}$$

The proof is completed. □

As a corollary of Lemma 2.1, we obtain the following.

Proposition 2.3. *If a group G of isometries satisfies* (G.1), *then the restriction $Q|X_G\backslash\{0\}$ of the functional Q on $X_G\backslash\{0\}$ satisfies the Palais–Smale condition.*

3. Proof of Main Theorem

To prove our main result we need the following.

Proposition 3.1. *Let X be an infinite–dimensional closed subspace of $H^1(M)$. Assume that the restriction $Q|X\backslash\{0\}$ of Q on $X\backslash\{0\}$ satisfies the Palais–Smale condition. Then, $Q|X\backslash\{0\}$ has a sequence $\{u_j\}$ of critical points such that $Q(u_j) \longrightarrow \infty$ as $j \longrightarrow \infty$.*

This result is proved by the mountain–pass theorem of Ambrosetti–Rabinowitz. For the proof, the reader is referred to [AR], [R].

We now give a proof of Main Theorem.

Proof of Main Theorem. Take the subspace X_G defined by (2.1). From (G2) we can easily see that X_G is infinite–dimensional. Then, we apply Proposition 2.3 and 3.1 to obtain the sequence $\{u_j\}_{j=0}^{\infty}$ of critical points of the restriction $Q|X_G\backslash\{0\}$ satisfying $Q(u_j) \longrightarrow \infty$ as $j \longrightarrow \infty$. Furthermore, using the direct method in the calculus of variations we may take u_0 so that

$$Q(u_0) = \inf\{\, Q(u) \mid u \in X_G\,,\ \ u \not\equiv 0\,\}\,,$$

holds. We remark that the functional Q is invariant under the action of isometries of (M, g). By the Palais symmetric criticality principle [Pa], if G is a compact Lie group, then any critical point of $Q|X_G\backslash\{0\}$ is also a critical point of Q. Thus, we see that each u_j satisfies (1.1) weakly with $\lambda = E(u_j)/\|u_j\|_N^N$. By the regularity theorem of Trudinger [Tr], we get $u_j \in C^2(M)$ for any j. Thus, we prove statement (1)—(3).

Finally, we prove statement (4). Since $Q(u) = Q(|u|)$ for any $u \in H^1(M)\backslash\{0\}$, we may assume that u_0 is non–negative everywhere. By the maximum principle and the regularity theory of elliptic equations, we observe that u_0 is a positive smooth function. The proof is completed. □

Remark 3.2. Schoen [S3] obtained the uniform estimate for any positive solution of (1.1) in case (M, g) is not conformally equivalent to the sphere. As a corollary of his result, we are able to show that for any positive critical point u the value $Q(u)$ is bounded above by a constant depending only on n, g. Thus, we observe that the critical point u_j obtained in Main Theorem changes the sign except finitely many j.

Example 3.3. Let M be the sphere S^n with the standard metric g_0. Take G as

$$G = O(k_1) \times \cdots \times O(k_m) \qquad \text{where} \quad k_1\,, \cdots, k_m \geq 2\,, \quad k_1 + \cdots + k_m = n + 1\,.$$

We easily see that G satisfies (G1) and (G2). Then, we get a sequence $\{u_j\}_{j=0}^{\infty}$ of G–invariant critical points of Q. In case $m = 2$, this result was proved by Ding [D]. On the other hand, Obata [Ob] proved that the minimizers are only the positive critical points.

Example 3.4. Let M be the product manifold $S^1(T) \times S^{n-1}$ of a circle of length T with a unit $(n-1)$–sphere. Let g be the product metric. We set

$$G_1 = \left\{ A = \begin{pmatrix} I_2 & O \\ O & B \end{pmatrix} \,\middle|\, B \in O(n) \right\} \cong O(n),$$

Schoen [Sc2] proved that any positive critical point is invariant under the action of G_1 using the method of Gidas–Ni–Nirenberg [GNN]. He described all positive critical points by the analysis of an ordinary differential equation. Since G_1 satisfies (G1) and (G2), we get infinitely many G_1–invariant critical points of Q which change the sign. We next prove that there exist infinitely many critical points which are not G_1–invariant. We set

$$G_2 = \left\{ A = \begin{pmatrix} B & O \\ O & I_n \end{pmatrix} \,\middle|\, B \in SO(2) \right\} \cong SO(2) \cong S^1,$$

Then, we see that G_2 also satisfies (G1) and (G2). We apply Main Theorem to get infinitely many G_2–invariant critical points. Since only a constant function is invariant under the action of both G_1 and G_2, we obtain the desired result.

4. Boundary Value Problem

Let Ω be a bounded domain in $\mathbb{R}^n$ ($n \geq 3$) having the smooth boundary. We here consider the nonlinear eigenvalue problem

$$\text{(4.1)} \qquad \begin{cases} -\Delta u = \lambda |u|^{N-2} u & \text{in } \Omega, \\ u = 0 & \text{on } \partial\Omega. \end{cases}$$

By the similar way to the Yamabe problem, we observe that (4.1) is the Euler–Lagrange equation of the functional

$$F(u) = \int_\Omega |\nabla u|^2 \, dx \Big/ \Big(\int_\Omega |u|^N \, dx \Big)^{2/N} \qquad \text{for} \quad u \in H_0^1(\Omega) \backslash \{0\}.$$

From the best constant of the Sobolev inequality (for example, see [A2]) , we deduce

$$\sigma := \inf\{ F(u) \mid u \in H_0^1(\Omega),\ u \not\equiv 0 \} = \frac{n(n-2)}{4} \operatorname{vol}(S^n)^{2/n}.$$

Pohozaev [Po] proved that there is no positive solution of (4.1) if Ω is star–shaped. From his result we can easily derive that the infimum σ is never achieved for any bounded domain Ω. We therefore see that any solution of (4.1) is a non–minimizing critical point of F. On the other hand, Bahri–Coron [BC] proved the existence of a positive solution of (4.1) under some condition on the topology of Ω.

We apply the same argument in section 2 and 3 to have the following.

Theorem 4.1. *Let G be a compact Lie group of $O(n)$. Suppose that Ω is invariant under the action of G and G satisfies* (G.1) *with $M = \Omega$. Then, there exists a sequence $\{u_j\}_{j=0}^{\infty} \subset C^2(\overline{\Omega})$ satisfying*

(1) *each u_j satisfies (4.1) with $\lambda = \|\nabla u_j\|_2^2 / \|u_j\|_N^N$.*
(2) *each u_j is invariant under the action of G,*
(3) $F(u_j) \longrightarrow \infty$ as $j \to \infty$,
(4) *u_0 is a positive smooth function which attain the minimum of F over G–invariant functions vanishing on $\partial\Omega$.*

Remark 4.2. Fortunato–Jannelli [FJ] proved the similar result in case $n \geq 4$ and Ω has *rotational symmetry*. Their proof differs from us.

Appendix

Here we give a proof of Theorem 2.2 following the method of Evans in [Ev].

Since $\{\nabla u_j\}$ is bounded in $L^2(M)$, we may assume that $|\nabla u_j|^2 dV$ converges to some measure μ weakly.

We first consider the case $u \equiv 0$. By the Sobolev inequality we have

$$\text{(A.1)} \qquad \Big(\int_M |\phi u_j|^N dV\Big)^{1/N} \leq C\Big[\Big(\int_M |\nabla(\phi u_j)|^2 dV\Big)^{1/2} + \Big(\int_M |\phi u_j|^2 dV\Big)^{1/2}\Big],$$

for any $\phi \in C^\infty(M)$ where C is a constant depending only on n.

Since $u_j \longrightarrow u \equiv 0$ weakly in $H^1(M)$, we deduce

$$\text{(A.2)} \qquad \Big(\int_M |\phi|^N d\nu\Big)^{1/N} \leq C\Big(\int_M |\nabla\phi|^2 d\mu\Big)^{1/2}.$$

By approximation, we have

$$\text{(A.3)} \qquad \nu(E)^{1/N} \leq C\mu(E)^{1/2} \qquad \text{for any Borel set } E \subset M\,.$$

Since μ is a finite measure, the set

$$\mathcal{S} = \{\, x \in M \mid \mu(\{x\}) > 0 \,\}\,,$$

is at most countable. We can therefore write $\mathcal{S} = \{x_i\}_{i \in I}$, $\mu_{x_i} = \mu(\{x_i\})$ so that

$$\mu \geq \sum_{i \in I} \mu_{x_i} \delta_{x_i}\,.$$

Since (A.3) implies $\nu \ll \mu$, we have

$$\text{(A.4)} \qquad \nu(E) = \int_E D_\mu \nu \, d\mu \qquad \text{for any Borel set } E\,,$$

where

$$\text{(A.5)} \qquad D_\mu \nu(x) = \lim_{r \downarrow 0} \frac{\nu(B(x,r))}{\mu(B(x,r))} \qquad \text{for} \quad \mu\text{–a.e. } x \in M\,.$$

From (A.3), on the other hand, we have

$$\frac{\nu(B(x,r))}{\mu(B(x,r))} \leq C^N \mu(B(x,r))^{2/(n-2)} \qquad \text{provided} \quad \mu(B(x,r)) > 0\,.$$

Thus we obtain

$$\text{(A.6)} \qquad D_\mu \nu(x) = 0 \qquad \mu\text{–a.e. on } M \backslash \mathcal{S}\,.$$

Now we define $\nu_{x_i} = D_\mu \nu(x_i) \mu_{x_i}$. Then, by (A.4) and (A.6) we obtain

$$\nu = \sum_{i \in I} \nu_{x_i} \delta_{x_i}\,.$$

We next consider the case $u \not\equiv 0$. We set $v_j = u_j - u$. From the theorem of Brezis–Lieb [BL], we have

$$|\nabla v_j|^2 dV \longrightarrow \mu - |\nabla u|^2 dV\,,$$
$$|v_j|^N dV \longrightarrow \nu - |u|^N dV\,.$$

Then, we obtain the desired result by applying the above argument to $\{v_j\}$. The proof is completed.

REFERENCES

[**AR**] A. Ambrosetti and P. Rabinowitz, *Dual variational methods in critical point theory and applications*, J. Funct. Anal. **14** (1973), 349–381.

[**A1**] T. Aubin, *Équations différentielles non linéares et problème de Yamabe concernant la courbure scalaire*, J. Math. Pures Appl. **55** (1976), 269–296.

[**A2**] ———, *Nonlinear Analysis on Manifolds. Monge–Ampère Equations*, Springer Verlag, New York, 1982.

[**BC**] A. Bahri and J. M. Coron, *On a nonlinear elliptic equation involving the critical Sobolev exponent : the effect of the topology of the domain*, Comm. Pure Appl. Math. **41** (1988), 253–294.

[**BL**] H. Brezis and E. H. Lieb, *A relation between pointwise convergence of functions and convergence of functionals*, Proc. Amer. Math. Soc. **88** (1983), 486–490.

[**D**] W.-Y. Ding, *On conformally invariant elliptic equation on* $\mathbb{R}^n$, Comm. Math. Phys. **107** (1986), 331–335.

[**Ev**] L. C. Evans, *Weak convergence methods for nonlinear partial differential equations*, regional conference in math. no. 74, Amer. Math. Soc., Providence, Rhode Island, 1990.

[**FJ**] D. Fortunato and E. Jannelli, *Infinitely many solutions for some nonlinear elliptic problems in symmetric domain*, Proc. Royal Soc. Edinburgh **105A** (1987), 205–213.

[**GNN**] B. Gidas, W.-M. Ni and L. Nirenberg, *Symmetry and related properties via the maximum principle*, Comm. Math. Phys. **68** (1979), 209–243.

[**LP**] J. M. Lee and T. M. Parker, *The Yamabe problem*, Bull. Amer. Math. Soc. **17** (1987), 37–91.

[**Ob**] M. Obata, *The conjectures on conformal transformations of Riemannian manifolds*, J. Diff. Geom. **6** (1972), 247–258.

[**Pa**] R. Palais, *The principle of symmetric criticality*, Comm. Math. Phys. **69** (1979), 19–30.

[**Po**] S. Pohozaev, *Eigenfunctions of the equation* $\Delta u + \lambda f(u) = 0$, Soviet Math. Dokl. **6** (1965), 1408–1411.

[**R**] P. Rabinowitz, *Minimax methods in critical point theory with applications to differential equations*, regional conference in math. no. 65, Amer. Math. Soc., Providence, Rhode Island, 1986.

[**S1**] R. Schoen, *Conformal deformation of a Riemannian metric to constant scalar curvature*, J. Diff. Geom. **20** (1984), 479–495.

[**S2**] ———, *Variational theory for the total scalar curvature functional for Riemannian metrics and related topics*, Springer Lecture Notes in Math. **1365**, 120–154.

[**S3**] ———, *On the number of constant scalar curvature metrics in a conformal class*, preprint.

[**Tr**] N. S. Trudinger, *Remarks concerning the conformal deformation of Riemannian structures on compact manifolds*, Ann. Scuola Norm. Sup. Pisa **22** (1968), 265–274.

[**Y**] H. Yamabe, *On a deformation of Riemannian structures on compact manifolds*, Osaka Math. J. **12** (1960), 21–37.